Investigative Pathways

Investigative Pathways

Patterns and Stages in the Careers

of Experimental Scientists

Frederic Lawrence Holmes

Yale University Press

New Haven & London

Printed in the United States of America by Sheridan Books, Ann Arbor, Michigan.

Library of Congress Cataloging-in-Publication Data

Holmes, Frederic Lawrence.
    Investigative pathways : patterns and stages in the careers of experimental
scientists / Frederic Lawrence Holmes.
        p.   cm.
    Includes bibliographical references and index.
    ISBN 0-300-10075-2 (alk. paper)
    1. Creative ability in science.   2. Discoveries in science.   I. Title.

Q172.5.C74H65 2004
501'.9—dc22

                                                                      2003057640

A catalogue record for this book is available from the British Library.
The paper in this book meets the guidelines for permanence and durability of the
Committee on Production Guidelines for Book Longevity of the Council on
Library Resources.

10   9   8   7   6   5   4   3   2   1

For Petra

Contents

Preface

When I entered the newly emerging academic discipline of the history of science nearly half a century ago, one of its central tasks appeared to be to account for the origins of the great discoveries on which the modern sciences were established. The views of that time were forcefully represented in René Taton's *Reason and Chance in Scientific Discovery,* the preface of which began with the declaration, "All scientific progress is the result of a chain of discoveries of differing degrees of importance and significance."

Historians of science then were aware that published accounts of discoveries normally did not reveal how scientists had arrived at the conclusions their publications were intended to support. There was, consequently, a special interest in locating and interpreting unpublished documents, such as correspondence or informal notes that the scientists may have kept, from which some of the steps along the way to the discoveries might be recovered. In the cases of many of the landmark advances of early modern times, such as Galileo's laws of motion or Lamarck's idea of the transformation of species, such surviving records have been sparse enough to leave much room for speculative historical reconstruction. In other cases, such as the work of Isaac Newton or Charles Darwin, the records left behind have been so dense that the reconstruction of the thought processes involved has been challengingly complex.

Persuaded that the history of science was preeminently a history of scientific thought, historians of that era gave priority to the origins of novel scientific ideas. They were, however, also interested, to a greater extent than their successors sometimes acknowledge, in the stages leading to the great experimental advances of the past. Among the canonical experimental figures was the French physiologist Claude Bernard, whose brilliant discoveries during the 1840s and 1850s were seen as laying the foundations for modern experimental physiology. In *An Introduction to the Study of Experimental Medicine,* Bernard himself described the origins of his major discoveries as illustrations of forms of experimental reasoning that he believed essential to the successful pursuit of his science. These accounts were taken to be deeply revealing of Bernard's experimental practice; but when the historian of medicine Mirko D. Grmek began during the 1960s to examine Bernard's laboratory notebooks, nearly all of which have been preserved, he found that for each of the discoveries that Bernard described, the actual circumstances were more complex than Bernard had portrayed them in his retrospective descriptions, and that in almost every case there were some contradictions between his recollections and the actual course of events leading to the discovery.

With Grmek's encouragement I began in the late 1960s to examine Bernard's laboratory notebooks with the initial aim of uncovering the earliest traces of his ideas about the internal environment. This effort proved impracticable, and I soon turned instead to chart his course in the study of the chemical processes of life from the beginning of his career in 1843 until he reached his first major discoveries in the field, in 1848. Prior to those successes, Bernard published about a dozen papers claiming discoveries of a relatively minor nature or asserting theories of digestion that did not long survive scrutiny. As I followed his progression from one to another of these publications, I began to perceive that one could not divide his experimental activity into discrete episodes constituting the steps leading to each of these discovery claims. Rather, the discoveries themselves were only markers along the way, sometimes merely by-products, of a more continuous, more comprehensive experimental quest to understand the chemical processes that foodstuffs undergo from the time they enter an organism until they are finally excreted. The patterns of this broader, persistent activity I began to view through the metaphor of the investigative pathway. Since then I have examined in similarly fine detail the extended experimental pursuits of several other highly successful scientists from the mid-eighteenth to the mid-twentieth century, whose daily laboratory records have also survived, and have found that the pathway metaphor is generally helpful in the reconstruction of their investigative lives.

The object of the present volume is to draw on my experience with these individual historical reconstructions in order to suggest some generalizations about the nature of sustained individual experimental research, its relation on the one hand to the purposeful organization of a creative life, and on the other to the broader development of the field within which the individual participates. Because these generalizations are grounded mainly in the particular scholarly pathway through which my own career has led me, I must stress that the resulting perspectives are necessarily personal and partial. Historical studies of the nature of scientific change must be pursued on many scales, from that of the daily operations of a single investigator working alone, to the effort directed by the head of a laboratory or research school in which several to many individuals participate, to the networks of one hundred or more investigators who make up the frontier of a specialty field at a given time, to the movements of a large branch of science over the span of decades or centuries. Individual research pathways underlie these larger-scale activities also, but as the scope of the activity to be described increases, the level of resolution at which it can be followed in a historical narrative necessarily diminishes. When we focus on individual investigative pathways, therefore, we are constrained to limit the scope of our studies to the finer of the many scales of scientific advance. As we move up through these scales of activity, individual pathways must be compressed, until at last we can afford to describe only the highlights of their outcomes, the contributions of each scientist to the progressive development of a collective field of activity.

Historians have long debated whether, and to what extent, it is possible to generalize from specific cases. We are acutely aware of the ease with which we can impose patterns on events, how prone we are to mistake for the lessons of history our presuppositions about the underlying meanings of the stories we tell. Nevertheless, few of us are satisfied to think that each story is so unique that it has nothing to say to us about similar stories that took place in different times and places, involving different casts of characters. While writing the narratives of the several scientists whose lives and work I have studied in close detail, I often paused to insert reflections on what seemed to me illustrative of scientific investigation in general. To make such casual asides, however, is far easier than to try to knit them together into extended discussions about the generalizations

common to more than one example. In writing this book, I have tried to balance generalization with narrative illustration, conscious that this balance is
never secure; that one necessarily walks a subtle line between the danger of fitting the cases too closely into the template one wishes to present, and that of losing the contours of the template in the complexities of the particular cases.

A reviewer of the manuscript noted perceptively: "Underlying many of the
chapters seems to be a preoccupation with continuity in science (at least when
one focuses on the level of an individual's life work) [but] the issue of continuity versus discontinuity" is not discussed in these terms. "I felt that the book was
an argument . . . for seeing connections and continuity where a more distant
view might suggest rupture."

There may be in current historical studies a trend toward seeing continuity
where writings of two or three decades ago stressed discontinuity. My purpose,
however, has not been so much to display continuity in place of discontinuity
as to show their complementarity. Developments that appear at a fine-grained
scale continuous often produce cumulatively outcomes that are, on a larger
scale, genuinely discontinuous with their starting points.

The organization of the material necessitates returning repeatedly to the same
events from different viewpoints and at different degrees of resolution. Some
repetition is unavoidable. Because I expect that some readers may selectively focus on certain chapters, I have allowed some details to reappear in a manner that
will seem redundant to the reader who has followed chapter by chapter, but also
allows each chapter to be read semi-independently of those preceding it.

Acknowledgments

Because this book integrates much of my previous experience as a historian of science, it is an appropriate place to express my heartfelt thanks to all those who have helped, sustained, and encouraged me both before and during the forty-five years since I entered that field. I cannot mention them all individually, and do not want to omit some of them by singling out others, but I do remember. They begin with my father and mother, who always made me feel that I could do something significant in my life.

A little more than a year ago, my dear friend and partner, Petra Werner, persuaded me that I should write this general book as soon as I could, not postpone it until I had finished yet another case history. Fortunately, I have taken her advice in time.

During a time when I needed it most, my three daughters, Catherine, Susan, and Rebecca, have been wonderfully supportive. I am especially grateful to Becky, who postponed her medical residency for a whole year to care for me. I hope and trust that her generosity of spirit will be long rewarded.

It has been a great pleasure to work, for the second time, with Jean Thomson Black and the editorial and production staffs of Yale University Press.

Introduction

In a study of creative genius, Dean Keith Simonton has described the "products" of outstanding individuals as separate "creative acts."[1] Howard Gruber, in contrast, sees a "creative process" resembling sustained growth. "Thinking about complex subjects," Gruber has written, "is *organized over time,* over long periods of time." The creative thinking person may appear to be wholly occupied at a given time by a sub-group of problems, but these tend to be held together, in episodic succession, by their part in a "network of enterprise" through which the individual expresses a prolonged sense of purpose.[2]

Such temporal continuity applies to creative activity in general, but with particular force in science. As John Ziman has noted, scientific work is, by its very nature, highly specialized. "The practical expertise and understanding needed to undertake a serious scientific investigation usually takes many years to acquire." Once an individual has attained such specialized experience, Ziman shows, he or she will most likely continue to work in that same area. "Given the freedom to do so, the natural tendency of most scientists is to concentrate for years on a few problems in a narrow area of research," sometimes over the entire span of a professional life. Some individuals "drift" or "migrate" across specialty boundaries or "diversify" into more than one area, but unless scientists are forced to change because of policies imposed by the organizations within which they work, such movements are typically quite gradual transitions rather than abrupt shifts. Ziman discusses these aspects of scientific careers under the heading "research trails," without commenting on the metaphorical significance of the spatial images he applies to temporal pursuits.[3]

The metaphor of the research trail—or the alternative phrase that I have adopted, the "investigative pathway"—both describes and can contribute to our understanding of the personal trajectories of individual scientists within the larger investigative movements in which they take part. The double face of each branch of science—as, on the one hand, a collective "long march" by a group of specialists sharing a discipline and pushing ahead together along a "frontier," and, on the other hand, as the private struggle of each individual within that group to find a distinct place and to produce original discoveries, conclusions, or other claims through which he can make recognized contributions to the movement—provides the creative tension around which all our stories about the "progress" of science somehow revolve.

The collective movement of science is open-ended, continuous through time for as long as such activity retains social support for the resources it requires. The individual trajectory is limited by the span of a lifetime and conditioned by the stages of a personal life. The opportunities afforded the individual to contribute to the larger development are not a function of intellectual capacity alone, but of temperament, training, prior experience, and the other changing aspects of life through adolescence, maturity, and old age. Individual investigative pathways thus display characteristics drawn both from the nature of the collective advance of science and from the personal needs of creative, purposeful, but mortal lives.

The pathway metaphor provides images through which we can express some of these features. Understood not as a preexisting, well-traveled route that the investigator *follows,* but as one that she *creates* while exploring territory previously untraveled, the pathway suggests that one proceeds step by step, each step guided by those taken previously and by uncertain intimations about what lies ahead. One is not normally led in a straight line toward a preordained destination. The investigator sets out with destinations clearly or vaguely in mind, but may shift direction, either because she encounters an obstacle she must go around, or realizes that her initial course will not lead where she expected it to, or because new goals emerge that divert her temporarily or permanently from her original intentions.

The pathway changes direction, but does not lose continuity. One can sometimes take shorter, sometimes longer steps, but in one step one can move only as far as one's stride allows. The metaphoric stride can, perhaps, be stretched farther than the literal one, but the investigator is rarely able to leap from the region in which she finds herself to a region within her discipline that is not in some respects contiguous to the one from which she moves.

The pathway metaphor is not merely imposed afterward by an outside observer of scientific careers. Scientists themselves frequently describe their personal experiences in ways suggesting that they envision themselves traveling along such trails. Late in his career, the eminent nineteenth-century scientist Hermann Helmholtz expressed his investigative ventures through the metaphor of searching for the way to the top of a mountain: "I would compare myself to a mountain-climber who, not knowing the way, ascends slowly and toilsomely and is often compelled to retrace his steps because his progress is blocked; who, sometimes by reasoning and sometimes by accident, hits upon signs of a fresh path, which leads him a little farther; and who finally, when he has reached his goal, discovers to his annoyance a royal road on which he might have ridden up if he had been clever enough to find the right starting point at the beginning. In my papers . . . I have not, of course, given the reader an account of my wanderings, but have only described the beaten path along which one may reach the summit without trouble."[4] Claude Bernard likened himself to a runner who chose not to follow the beaten path: "Everyone follows his own path. Some have been trained for a long time and proceed by following the track that had been marked out. I myself have reached the scientific arena by indirect ways and have been freed from rules by running forth cross-country, which others perhaps would not have dared to do. But I believe that in physiology that has not been bad, because it has led me to new views."[5] The American biochemist Vincent du Vigneaud delivered in 1950 a set of lectures retracing the "research trails that I have had the pleasure of working out in association with various collaborators over a period of twenty-five years." It is intriguing, he wrote, "how one starts out on a trail of exploration in the laboratory not knowing where one is eventually going, starting out, to be sure, with some immediate objective in mind, but also having a vague sense of something beyond the immediate objective, toward which one is striving. True exploratory research is really the working out of a winding trail into the unknown."[6]

Some of the British scientists who responded during the 1980s to John Ziman's queries about their research trajectories similarly resorted to pathway metaphors: One said, "I think the ways in which I would like the research to go have always been fairly obvious, in that I haven't come across any obvious crossroads wondering which way to go: it's always been quite clear which way to go." Another replied, "we are in a field where there is no sudden dead end. I mean, normally, research evolves, you know, in various directions, and you can, you know, if you see a dead end at one end, you jump to another branch; the whole thing is progressing in the same direction."[7]

The investigative pathway may be, at its most basic level, only the scientist's version of the metaphor of life as a journey, a manifestation of the most prevalent and inescapable of all metaphorical transformations, the representation of mental activity through spatial language.[8] Nevertheless, the capacity of this metaphor to differentiate particular patterns of movement within the general concept of the journey makes it a useful tool for probing the course of a scientific life.

That scientists typically spend many years pursuing one problem or a small set of related ones entails some drawbacks. Fields of science move quickly, and the trajectory of an individual is often overtaken by the collective advance. Rarely can a scientist remain indefinitely in the forefront, even of a field that his own contributions have opened up. What has been the leading edge of a frontier quickly becomes the settled territory of a "mature" field. In consolidating his own earlier work the individual often ceases to be seen as a leading figure in the broader field, or the field itself ceases to command the attention it once did. What we often see as the decline of creativity in the aging scientist may be more realistically understood as the disparity between the pace of change in science as a whole and the pace at which the individual can pursue the agenda set by early successes. It would appear on the surface that a scientist ought more easily to attain multiple successes by abandoning earlier pathways and making the personal leaps necessary to begin new pathways, or to join those more recently opened up by others. Occasionally, extraordinarily talented or energetic individuals are able to do this, but the rarity of such patterns compared to those in which the individual persists along the path set by his own previous work, or drifts or migrates gradually toward another field, suggests that there must be strong reasons for proceeding from where one has personally arrived, rather than attempting to relocate somewhere else in the collective venture.

A simple explanation for the prevalence of continuous investigative pathways is the "minutely specialized" nature of scientific work.[9] Each individual scientist acquires cumulative experience that is lost if she switches to a field to which that experience does not apply. Scientific experience is acquired by long apprenticeship, normally under the guidance of a mentor who can lead the emerging scientist to a place near the forefront of a problem of current interest. To begin anew on another frontier requires another apprenticeship, this time under circumstances in which the scientist is assumed to be mature and independent. Learning new skills takes time, and there is the risk that the new field to which he aspires to contribute will already have moved on before he can reach the level of expertise he already enjoys on his own earlier research pathway. History records more than one prominent scientist who moved, with misplaced self-assurance, from a field in which long preparation had made him a leading figure, to another one in which the superficiality of his knowledge and the brevity of his experience made him prone to serious mistakes.

Such ready explanations are given greater force by recent studies on the "acquisition of expertise" which show that in many activities, from chess to composing music to theoretical science, it requires on the average about ten years to reach a level of performance at which one can become truly creative on an international scale.[10] More powerful still, I believe, is the analysis that Michael Polanyi has given of science as "personal knowledge." The skills that enable a scientist to perform successfully, Polanyi argued, are not only the forms of knowledge and technique that can be articulated, but also those, which he called "tacit knowledge," that enable one to do things without being fully able to specify how they are done. These skills are like the tools that we not only learn to use, but which become extensions of ourselves, so that we eventually use them without being fully aware of their presence. Our attention is on the objects that we manipulate with their assistance. To pay too much attention to the tools themselves only diverts us from the task at hand. Such tools, whether material or conceptual, "can be conceived as such only in the eyes of the person who *relies on them* to achieve or signify something. *This reliance is a personal commitment which is involved in all acts of intelligence by which we integrate some things subsidiary to the centre of our focal attention.* Every act of personal assimilation by which we make a thing form an extension of ourselves through our subsidiary awareness of it, is a commitment of ourselves, a manner of disposing of ourselves."[11] If the skills and experience a scientist accumulates while following an investigative pathway become his "personal knowledge," then to switch to another pathway requires more than a different set of experiences. In some deeper sense it involves a new personal identity.

Scientists commonly stick to their previously defined investigative pathways not merely because it is difficult to change to another unrelated one, but because to persist along the same one is in many cases a highly effective research strategy. That a lifetime can be productively spent in this manner reveals something about the nature of scientific problems that is obscured by our fascination with individual discoveries, "breakthroughs," the excitement of the latest scientific news, and the strong tendency of science to bury its own past. Seldom does a single experiment suffice to test a given hypothesis, settle a disputed point, or reveal a hitherto hidden regularity. Most often an investigator performs a long series of experiments that slowly close in on an unknown phenomenon, build up the case for, or dissolve the case for, a tentatively held idea. The problems whose solutions lead to publications and even to eminent personal distinction are most often sub-problems within the broader problems that define the goals of an investigative pathway. It is because nature is so constructed that very few important problems are solved all at once, that scientists can so often pursue a set of related problems year after year, without ever coming to the end of a well-conceived, productive research trail. Problems are nestled within problems, and many solutions are merely way stations along the route. Satisfying in themselves and worthy of recognition, they point the way to a further road ahead. New methods or concepts becoming available enable the researcher who has persisted for many years finally to begin to solve problems that had earlier defied her most resourceful efforts. Solutions at one level leave deeper questions unanswered. Sometimes the deeper questions elude the conceptual and material tools that can be brought to bear on them, and the research pathway is blocked. But seldom does it end simply because a solution leaves little more to do.

A special power of the concept of the investigative pathway lies in the conjunction between its effectiveness as a research strategy and its expression of the distinctiveness and continuity of the individual scientific personality. Scientists are, and have been for more than two centuries, constrained to fit themselves into modes of practice prevalent at the time they enter a given field. The older heroic image of the great scientist as one who possesses from the beginning more profound, unerring insight than those already active in the field, who takes up from an entirely original point of view problems that have stalled his predecessors, or that they have failed to notice as problems, can no longer be taken seriously. Were an individual to begin with complete independence, the likelihood is that he would solve problems uninteresting to the field, or produce solutions unrecognizable by his contemporaries as solutions. The results of his labors would not appear to others to be genuine discoveries.[12] Even the most talented of investigators normally begin by learning to practice within a field in ways that are currently in fashion, learning from the most successful recent exemplars how to go about producing further successes. As the individual moves beyond these initial steps in a new research trail, however, the accumulation of personal decisions about which way to move from where she is toward a promising point nearby gradually distinguishes her trajectory from those of her contemporaries. As the pathway continues to lengthen and take on new directions, it comes more and more to resemble the personal signature of a particular scientific life.

In this book I hope to illustrate and develop these generalizations in large part from my experience reconstructing in fine detail portions of the investigative pathways of several experimental scientists whose lives are spread out over three centuries. In these studies I have become aware, on the one hand, of the many personal differences of method, approach, and style that mark each scientist as both a unique individual and a participant in a unique period of time; and have sensed, on the other hand, some deeper underlying commonalities that link them as participants in the same grand quest, in which each endeavors to advance knowledge of some small part of the natural world. It has seemed to me that the more closely one penetrates to the day-to-day activities of investigative lives, the more similar, rather than the more different, do these lives appear, even though widely separated by time, by field, and by the nature of the tools available for the job.

These individuals I will take to be representative of a much larger class of creative scientists who have had sufficient personal success to count among the leaders of their particular fields during their lifetimes. Three of them have attained a historical stature that transcends their own eras. They are not, therefore, necessarily typical of the many more scientists who have labored competently in the vineyards of science without achieving unusual distinction. Nevertheless, their personal trajectories can illuminate patterns characteristic of investigative pathways that may apply also to the experiences of many of their colleagues.

I shall return repeatedly to the experiences of these individuals, because the reconstruction of the "fine structure" of portions of their investigative pathways reveals features that are less obvious for scientists whose work has not been studied in comparable detail. The continuity of investigative pathways is best displayed when we can trace the individual "steps" that make them up. Shifts that appear abrupt from a greater distance can then be seen to resolve into transitions the investigator has made by traversing some connecting branches between two main routes of her research trail. Some of the larger-scale patterns of investigative lives can, however, be explored without such access to the fine structure of their daily activity, and I shall draw comparisons where appropriate to the lives of other scientists whose work is known through biographical treatments of more ordinary length.

Part One Interpretations of Scientific Discovery and Creativity

Chapter 1 Investigation, Discovery, and Experimental Practice

In his study of the activities of a group of followers of William Harvey who pursued, during the decades between 1640 and 1690, a cluster of unsolved problems made prominent by Harvey's discovery of the circulation of the blood, Robert Frank wrote in 1980 that the older historians of science had tended to treat only the "products" of past developments: that is, the explanations reached for the phenomena examined. Yet, Frank pointed out, these historians were most often themselves practicing scientists. He wondered why such men "failed to read into past literature the dynamic qualities of conjecture and experiment, of question and response, of interaction with colleagues, that had dominated their own lives." His study was "conceived under the rather different assumption that the historian can, if the appropriate evidence survives, recapture even for a period as distant as the seventeenth century, that sense of process that animates the science of the recent past."[1]

Frank succeeded admirably in his purpose. Assembling a rich variety of primary published and archival documents, he was able to bring to life a community of natural philosophers, linked both by intersecting interests and by activities that intersected informally in the compact seven-teenth-century university town of Oxford. Individually and collaboratively they worked on problems, such as how respiration is carried out, the role of air and its properties, the formation of heat in animals, the properties and functions of the blood, and the motions of the heart. They asked questions, tried to answer their questions, developed new points of view, often failed to solve the problems they took up, but occasionally made important discoveries. To a degree unusual in the history of science, Frank was able to portray the way in which a collective research front is composed of a network of tightly interacting individual research trails.

Similar concerns motivated Martin Rudwick's "fine-grained" reconstruction of the scientific activities of a group of British geologists who took part, during the 1830s, in what became known as the Great Devonian controversy. The controversy began as a dispute over the local identification and sequence of strata in Devon, a coastal county in England, and ended in the establishment of a global geological system. Employing a remarkably complete network of surviving correspondence, Rudwick was able to follow the monthly, weekly, sometimes even daily work in the field of the several geologists who played major roles and the many who played minor roles, as well as the stages in the development of their arguments, as the controversy unrolled and led eventually to consensus. To achieve genuine understanding of the "*processes* by which new knowledge is shaped," Rudwick applied a strict rule never to refer at any point in his narrative to events that had not yet occurred or to thoughts that his participants had not yet had.[2]

These two books can serve as models for many more studies of the dynamics of individual and collective scientific investigation: models that examine the discoveries made along the way, but do not view them as the only significant outcomes of ongoing processes. The vast terrain of the modern scientific enterprise, as it has grown from the endeavors of the small groups of seventeenth-century natural philosophers to the massive scale of late twentieth-century research, presents innumerable favorable sites for such studies. The promise offered by these models has not, however, attracted the central attention of the field. Nor has the earlier focus on the origins of great discoveries retained the intense scholarly interest it once held for historians of science. The most prominent trends in the field have been in other directions, many of them of absorbing interest, but aiming mainly not at the dynamics of the scientific enterprise itself so much as at the connections between science and broader social or cultural landscapes.

Meanwhile, the territory left partially vacant by historians has been filled by other disciplines.

Under the formidable influence of Karl Popper, philosophers of science for several decades from the 1930s until the 1970s fixed their attention on the "context of justification," accepting Popper's dictum that discovery was a psychological phenomenon not subject to rational analysis. By the end of this period, however, some philosophers had come to reject the distinction between the contexts of discovery and justification, and to predict a time "in which discovery, innovation, and problem solving will take their places as a legitimate area of study."[3] To a modest degree this prescription has been followed. For example, Kenneth Schaffner has used several concrete examples taken from the recent biomedical sciences to support the "thesis that the process of scientific discovery involves logically analyzable procedures, as opposed to intuitive leaps of genius."[4] Lindley Darden has employed particular historical examples, such as the development of Mendelian genetics, to "explore the dynamics of theory change." Focusing on the theory of the gene, whose classical expression was completed by 1926, Darden shows how the component parts of the theory emerged gradually during the first quarter of the twentieth century, and she draws from this case a comprehensive list of "strategies of theory change" that might have generated the various components. Her analysis rests entirely on the published record of scientific thought, but the strategies she suggests might also be applied, for historical instances where suitable unpublished documents have survived, to examine the more intimate level of the generation of new ideas by individual scientists before they have reached the stage of publication.[5]

David Gooding has studied intensely the laboratory notebooks of Michael Faraday and repeated for himself some of the critical experiments of Faraday and his predecessors concerning the interactions between magnetism and electric currents, in order to understand the intimate encounters between experimental scientists and nature, the tacit and articulate knowledge applied by the experimenter to design and modify experiments, the ways in which experimentalists initially "construe" their experiences before constructing coherent theoretical explanations, the ways in which experimentalists transform perceptual experience into meaningful language, and many other aspects of the "play of actions and operations in a field of activity" that he calls the "*experimenter's space.*" By reconstructing the successive steps an experimenter takes in schematic, maplike form, Gooding represents the experimental pathway over time as a spatial pattern, and shows that such pathways are not linear but replete with "bypasses, recycling, repetition, [and] re-interpretation of earlier results." He uses the results of his study to challenge many of the tenets of a philosophy of science too exclusively oriented, in his view, around mental and verbal processes and too little attentive to the material aspects of scientific practices.[6]

More prominent and prevalent during the last two decades than the studies of experimentation by philosophers have been those of the "details of the day-to-day doing of science"[7] by a heterogeneous circle made up of sociologists, ethnomethodologists, anthropologists, and some historians who have taken their cues from these fields, who identify themselves collectively under the general heading of "science studies," or SSK. Tracing their lineage to the work of Thomas Kuhn, whom they regard as having opened the way, but who had traveled only half the distance from the "received view" of science as the objective search for truth to their own doctrine that scientific knowledge is "constituted" by social processes,[8] the science studies group has sought to replace the older, dynamic concept of scientific investigation as the engine of discovery and of progressive advance with the more static concept of "experimental practice."

A major early stimulus to such studies came from the book *Laboratory Life*, by Bruno Latour and Steve Woolgar, one of whom resided in the laboratory of Roger Guillemin at the Salk Institute for several months as an anthropologist seeking to understand the activities of the laboratory as the behavior of a "strange" tribe. Latour and Woolgar illuminated many aspects of everyday laboratory life. Their central contentions—that pieces of "reality" are "constructed and constituted through microsocial phenomena"; that "scientific activity is not 'about nature,' it is a fierce fight to *construct* reality"; and that the "*laboratory* is the workplace . . . which makes [such] construction possible"[9]—however, appear less as insights gained through their experience in this laboratory than as presuppositions they had brought into it with them and relied on to interpret what they saw there. Similar presuppositions structured other well-known studies by the science studies group, including the highly influential account of the experimental practices of Robert Boyle in Steven Shapin and Simon Schaffer's *Leviathan and the Air-Pump*.[10] These studies shared a point of view well articulated by Shapin in 1982. "Neither reality nor logic nor impersonal criteria of the 'experimental method,'" Shapin wrote, "dictates the accounts that scientists produce or the judgments they make." Natural reality does not possess the "coercive force" with which the discourses of participants seem to endow it. Rather, "Reality seems capable of sustaining more than one account of it, depending on the goals of those who engage with it."[11] What shall "count" as a good experiment or an adequate representation of a natural phenomenon is, therefore, according to followers of this view, not a matter of the closeness of fit to any reality existing beyond contingent beliefs of the moment, but the outcome of negotiations among "core groups" who settle such matters. Attention is shifted from the encounters between investigators and the piece of nature they hope to come to understand, to the process by which the investigator attains authority for the claims he makes about "knowledge" produced by his experimental practice.

The science studies group has added significantly to the dimensions of our awareness of the nature of experimental practice. Among other things, they have drawn attention to the close interactions between the performance of an instrument and the objects it purports to observe or measure, and to the change in the status of an instrument when it is transformed from an object of study itself to a trusted means to investigate other objects. They have explored meaningfully the significance of the physical organization and control of access to the places where experiments are performed.[12] In their own estimation, "Science studies has been an exciting field for the past few decades, and one source of this excitement has been a continual expansion of conceptions of science as an object of studies."[13] Some of the more recent trends in their writing suggest, however, that the assumption which most clearly differentiates their approach from that of more "traditional" historians of science—that is, that nature plays in scientific investigation, at most, a role deeply subordinate to the construction of knowledge by human agency—may have led the science studies project toward a cul-de-sac from which its advocates are seeking to escape in ways that edge them back closer to the "received view" whose rejection was their point of departure. Thus Andrew Pickering now finds room for both "material agency" and "human agency." The fact that experimentalists must often modify apparatus and equipment that fails initially to produce the desired results he attributes to "resistance" emanating from their encounters with this material agency. The process of "accommodation" to this resistance until the experimentalist is successful is called "tuning." Pickering finds now, in addition, that it is necessary to take into account human "intentions," and that time, too, is an essential factor in experimental practice. Scientists cannot fully control or foresee the outcomes of their conceptual and experimental practice, and time alters both their intentions and their performances.[14]

These are important insights concerning the nature of the scientific enterprise, but the novel language in which Pickering has framed his modifications of the earlier science studies agenda tends to obscure the fact that what he presents as a further advance in the development of that program appears to a historian of science as a return to more ordinary concepts of historical process. That original intentions are seldom fully realized, that when one enters a pathway, whether it be in a scientific or any other realm of human endeavor, one cannot foresee how it will end, that one must continually accommodate one's original designs and plans to outcomes that differ from expectations, are all everyday matters to historians, for whom the unpredictable effect of the passage of time is a premise of their craft.

That the science studies agenda has provided new dimensions to our understanding of the everyday practice of science is undeniable; but its claims to have defined the directions in which the greatest advances in this area have been achieved in recent years cannot be accepted without qualification by those who have studied the dynamic processes of scientific change from other perspectives. Pickering has graciously mentioned my work as the most "tenacious" pursuit of the theme of the day-by-day doing of science "within traditional history of science."[15] There is nothing traditional, however, about reconstructing investigative pathways from laboratory research records, just as there is not about the methods that Frank and Rudwick have applied to elucidate the dynamics of group investigative activity. The historians of science of the last two decades have too easily ceded the study of these processes to other disciplines or allowed these disciplines to set the framework within which such processes are interpreted. I hope that this book can provide an alternative vision of the nature of scientific investigation to encourage future historians of science to reclaim a domain of historical enquiry that ought to lie at the heart of our enterprise.

Not all historians of science have abandoned the effort to reach deeply into the historical understanding of the processes of scientific investigation. Among the works of those who have persisted, an outstanding example is Jed Buchwald's study of the early career and experimental progress of Heinrich Hertz. Buchwald has followed the development of both Hertz's experimental techniques and the theoretical structures underlying them during the period in which Hertz began under the strong intellectual dominance of his teacher, Hermann Helmholtz, gradually achieved his own intellectual independence, and finally devised the first instrument to produce electromagnetic radiation. Buchwald has constructed the narrative of these events at a level of detail and penetration that captures the intricate thought processes of these leading figures in late nineteenth-century physics.[16] This book has had an admiring but limited readership, because Buchwald has not been afraid to describe in its full complexity the reasoning of two very powerful thinkers concerning highly technical matters. To grasp the arguments requires a high level of mathematical literacy. This is a dilemma for a history of science that not only confronts the formidable task of understanding the investigative pathways of scientists such as Helmholtz and Hertz, but is challenged to communicate what it finds to a "general" audience often reluctant to enter into specialized territory. If we are truly to understand experimental practice, however, we cannot evade that problem, or be satisfied with surrogate solutions that would tell us that even the knowledge pursued or acquired by such towering intellects is merely constituted by social processes.

The interplay between ideas and experimental operations has continued to be a strong theme in the recent work of some distinguished senior scholars, including John Heilbron, Alan Rocke, and Mary Jo Nye. Gerald Geison's *Private Science of Louis Pasteur* has drawn most attention for its controversial exposé of the misrepresentations of his work that Pasteur gave in public, but what Geison revealed was made possible by his careful reconstruction from laboratory notebooks of critical stages in Pasteur's research pathway.[17]

Particularly encouraging is that a younger generation of historians of science appears to be returning to the themes that have been for so long subordinated to other trends. In *The Life of a Virus*, a history of the large role that one particular organism has played in the exploration of broader questions about the nature and roles of viruses in our understanding of basic biological processes, as well as in practical approaches to disease therapy, Angela Creager fulfills her assertion that "in order to recapture experimental dynamics of this kind, the historian must focus on research at close range." A major theme in her book is the investigative pathway of Wendell Stanley, whose crystallization of tobacco mosaic virus in 1935 made both TMV and himself icons for biomedical research during the next decades. In *Picture Control: The Electron Microscope and the Transformation of Biology in America, 1940–1960*, Nicolas Rasmussen has shown how the introduction of this powerful new instrument shaped the research of a number of crucial investigators during those years. Nathaniel Comfort's new biography of Barbara McClintock probes deeply into the development of her remarkably individual research pathway in maize genetics, as well as into her strongly defined personality.[18] In *Pavlov's Physiology Factory*, Daniel Todes has explored the manner in which Ivan Pavlov managed the daily activity of a large-scale physiology laboratory at the end of the nineteenth century, showing how it functioned in generating experimental data, and how Pavlov himself provided the vision that both dictated the nature of the experiments carried out by the many *Pratikanti* who worked there for relatively short periods, and interpreted the results to fit the goals of his sustained investigation of the processes of digestion.[19]

Further signs of a revival of interest among historians of science in the processes of scientific inquiry was a workshop held in 2000 at the Max-Planck-Institut für Wissenschaftsgeschichte in Berlin by a group of historians, each of whom has used research notebooks to reconstruct portions of the experimental or theoretical activities of scientists ranging from Galileo and Newton to Galvani, Ampère, Pavlov, Einstein, and Hans Krebs.[20]

These are all hopeful signs, but we have far to go. Modern science is a huge enterprise. To explore comprehensively how scientific investigation has been done and is done at all levels, from the intimate daily activity of each person involved, to the broad investigative streams formed by the collective work of all of those engaged in a major specialty field, is a task still before us. The opportunities for such reconstruction are so much more plentiful than are the number of historians engaged in it, that we have so far occupied only a few outposts within a vast territory.

Chapter 2 The Three Scales of the Investigative Pathway and the Materials for Their Reconstruction

The eminent French historian Fernand Braudel introduced the term *longue durée* to designate those periods, lasting often for many centuries, during which shorter-range human activities are played out within general boundary conditions set by structures that remain "semi-immobile." The most accessible of such constraints are geographical conditions, including climate, vegetation, animal populations, the locations of villages and trade routes. Stable economic, social, cultural, and mental structures, however, may similarly limit for centuries the range of human activity. Braudel maintained that there are several independent rhythms of history to which historians must attend. For the social and economic history with which he was most concerned, he defined the mid-level as that of oscillatory cycles of prices, economic activity, or national products that typically last from a decade to half a century. The study of the rhythms of these movements he called the "new conjunctural history." Short-term history consists of the events due to human actions with which traditional histories have most often been preoccupied. Braudel believed that these concepts could be applied also to the history of science, and offered as examples two successive periods of longue durée dominated, respectively, by the Aristotelian universe and by the geometric universe of Galileo, Descartes, and Newton.[1]

Although he argued most forcefully for the longue durée, as the aspect of history least studied by previous historians, the "sole error," he claimed, "would be to choose one of these stories to the exclusion of the others." Whatever period or type of history one studies, it is imperative to take account of the full "hierarchy of forces, currents, and particular movements," to consider them in their ensemble, to "distinguish at each instant of one's research between long movements and brief thrusts."[2] With some adaptive modifications of definition we might apply Braudel's insistence on attention to each of these scales of history to the examination of the several scales that make up the life of an individual scientific investigator. Howard Gruber has made a persuasive case for the necessity of studying such lives at different degrees of resolution if we are to achieve full comprehension of their activity. The stream of creative thought, he wrote, is incredibly swift, but the growth of a new point of view is long and slow. During that slow evolution the scientist may solve a succession of problems that arise along the way. The achievement of these solutions may appear to the thinking mind as moments of insight that break sharply with what came before, but, in fact, they probably represent only minor nodal points "like the crest of a wave in a long and very slow process." Between the extremes of the stream of thought and the slow growth of a point of view, a "creative life is also episodic—organized in temporally compact periods within which a given orchestration of effort is played out, and certain projects executed."[3]

Rudwick, too, has called for studies of how science is done at "every degree of temporal graininess." Between the extremes of the longue durée and the experimental psychologist's study of scientific thinking on a scale of seconds or minutes "lies what is perhaps the most promising level of description to which the historian can hope to contribute: namely, the level of reconstruction plotted . . . in years, months, and weeks, and—with a bit of luck—sometimes even in days. This is, as it were, the high quality 'light microscope' of the analysis of scientific research practice. It shows somewhat less detail than the 'electron microscope' of still more fine-grained studies, but what it does show is less confusing and much easier to relate to the larger-scale of long-term features that can be seen with the . . . 'naked-eye' studies of conventional historical analysis."[4] The level of graininess most revealing of the patterns of scientific investigation may vary considerably depending on the nature of the historical problem studied, and whether, as in Rudwick's case, the research traced is that of a group, or is that of an individual. For the latter, I have found that the "fine structure" of experimental investigation can often be reconstructed meaningfully at the level of the succession of individual experiments, each of which may last from as little as half a day to several weeks, depending on the length of time required to prepare and perform them. The stream of thought of the investigator may, as Gruber suggests, be swifter still, but the progression of the thoughts of a laboratory scientist is normally constrained by the time it requires to subject the next step in his reasoning to an experimental check.

It is often not possible to tell stories simultaneously at several levels of resolution. To some degree, recapitulation can bring out larger features of a narrative obscured in the original telling by the plethora of details necessary to portray the finer-grained episodes and events that make them up; but the fine-grained story itself cannot easily be sustained over long enough periods to fill the entire narrative structure of larger-scale events. Frank was able to extend his account of the Oxford followers of Harvey over "several generations," but his ability to do so was partly due to the fact that the activity itself was not intensely maintained through the entire forty years from the first contacts between group members to their dissipation during the 1680s. The main events of Rudwick's Devonian controversy fortunately played themselves out within a time span of less than ten years. In my studies of the investigative pathways of individual scientists, the longest time span I have been able to sustain—the activities of Hans Krebs from the time of his apprenticeship to Otto Warburg to the discovery of the Krebs cycle—lasted ten years and occupied two lengthy volumes. Finely grained studies can seldom be expected, therefore, to cover even the lifetime pursuits of individual scientists, let alone extended periods in the collective advance of a research front. They can, nevertheless, illuminate the larger narratives if we make the reasonable assumption that what we find out from them about the nature of scientific activity is representative also of what we cannot cover within their circumscribed boundaries.

The degree of detail and resolution we can reach in the reconstruction of investigative pathways is limited mainly by the quality and quantity of the original documentary evidence and by the accidental circumstances that determine how much of it has survived. Like other people, scientists differ greatly in the care with which they make and retain records of their daily activities. It has been customary, from at least the seventeenth century, for experimental scientists to record in some systematic way the results of their laboratory operations. Such records served immediate practical purposes, such as allowing the investigator to review what has been done in order to know what still needs to be done, and to provide the material that will be presented in subsequent publications. Fortunately for the historian, such records have often been kept in bound notebooks, and the individual entries frequently dated. Such records preserve the most essential feature of an investigative pathway, the order in which things were done. They provide the chronological structure for a historical reconstruction, a backbone along which supplementary information from other types of supporting documents can be fitted.

In the most sparse cases, such research notebooks may contain little more than numerical results, together with minimal descriptions of experimental conditions. For their own purposes, scientists typically need not record their motives for carrying out the operations recorded, their expectations, or their responses to the outcomes of given trials. Even in such cases, the historian can often infer what is not directly recorded. The reason for doing a particular experiment, and perhaps the outcome to be anticipated, may appear quite obvious when the preceding research pathway has been reconstructed, while the response to the actual result may be best expressed in what the experimentalist tries next.

Sometimes we are luckier, however, and come across notebooks in which the scientist has spontaneously put down such intentions and responses. The laboratory notebooks of Antoine Lavoisier often are in the form of little narratives, in which he described not only the course of the experiment and its numerical results, but unexpected turns of events, some of them having the effect of ruining his efforts, others alerting him to new opportunities. Nor was he averse to noting his emotional responses, such as anxiety over a possible leak in his apparatus or disappointment over the failure to achieve the goal of an operation. Claude Bernard also wrote down narratives of experiments, sometimes while they were still under way, sometimes just after their completion. Impulsive and less than systematic, he sometimes left only cursory descriptions, at other times much fuller ones, and his descriptions sometimes merged into stream-of-conscious reflections on the outcome, on ideas to explain a result, or on what he ought to do next in light of the result. These entries offer deep insights into the thoughts of a creative scientist at work, but the historian cannot count on finding such information at key points. Often at places within Bernard's notebooks where the situation suggests that he must have reached a momentous threshold, there are no comments to confirm his awareness of the significance of the event. In such cases the historian must rely on his own judgment that he has identified such an event from the logic of the situation.

The experimental journals kept by Michael Faraday are systematic and richly revealing. In addition to full records of what he did, they contain diagrams of experimental apparatus that show us how he visualized what he intended to set up, and often how he visualized the phenomena he observed less directly. Because Faraday's experimentation on topics such as electromagnetism was also of extraordinary significance for the history of physics in the nineteenth century, these notebooks have attracted much attention from historians and philosophers of science, who have used them, or portions of them, to analyze the intimate interplay between thought and action in the work of one of the most imaginative of all experimental scientists.[5]

The great French scientist Ampère was much less systematic about record keeping. When Friedrich Steinle recently sought to reconstruct the short but critical research pathway during which Ampère studied the newly discovered effects of an electric current on a magnetized needle in the fall of 1820, he found in Ampère's extensive *Nachlass* only scattered, undated pages that appeared to be notes on experiments to be performed, or their results, that could be identified with this period. With considerable ingenuity, Steinle first "reconstructed" a laboratory record from these notes, then built further upon that reconstituted record an interpretation of the investigative pathway itself.[6]

Correspondence has long been regarded as an especially revealing source of information about the thoughts of investigative scientists in the course of their research. When it can be used in combination with laboratory records, the genres may supplement each other very productively. Letters seldom document the research step by step, but may contain explanations, interpretations, and expectations more fully developed than one is apt to find in the records an investigator keeps mainly for himself. In the case of collaborative investigations, or the interactions of scientists in collective activity and controversy, correspondence can become much closer to a primary record of events. This is particularly true, as Rudwick has pointed out, for the nineteenth century. Letters were delivered, at least through Great Britain and continental Europe, with a rapidity that is astonishing even by present standards. For individuals not in daily personal contact, letters were the primary, nearly exclusive means of communication. Nineteenth-century scientists often kept up a volume of handwritten correspondence that is today hard to comprehend.[7] The letters that geologists who participated in the Devonian controversy exchanged enabled Rudwick to follow in nearly seamless detail the course of the thoughts and work in the field of nearly every one of them.[8] Letters exchanged between the chemists Justus Liebig and Friedrich Wöhler during the periods in which they carried out collaborative experimental projects in their separate laboratories in Giessen and

Göttingen during the 1830s allow one to follow their nearly daily progress. Because each had to explain to the other what he had done and what he planned to do next, these letters are more revealing than most laboratory records, of the experimental reasoning each applied to the work.[9]

The nineteenth century may have been exceptionally favorable for the production and preservation of revealing correspondence between scientists, but the more recent past is less impoverished than one might suppose for an era in which the telephone is supposed to have superseded the letter as the preferred medium for short-term communication. The early molecular biologists of the 1950s, for example, formed an international network within which they maintained lively contact by letter, and seemed freer in reporting and discussing work-in-progress than is the case today. Some of this correspondence, such as that of Max Delbrück, has been collected in archives,[10] but much of it resides in the filing cabinets of scientists still active. The full extent of such material has, therefore, not yet been established, but the sampling that I and other historians of science working with selected individuals have done suggests that it may be as full and rich as for any other historical period.

In the course of drafting research papers, scientists often clarify or modify their interpretation of the results obtained, notice gaps that must be filled by further experimentation, and formulate or reject conclusions tentatively reached along the way. Far from an imperfect representation of what has already been done and thought, the composition of scientific papers is a further stage in the creative cycle, one which transforms portions of an open-ended research trail into bounded investigations with starting points and outcomes. The archives of Antoine-Laurent Lavoisier contain several successive drafts of some of his most important publications, affording a rich opportunity to follow a creative mind at work in this stage of the investigative process.

Laboratory records, letters, and drafts of papers can, with luck, be found for any period within the last three centuries. During the twentieth century some additional forms of documentation emerged. The systematic financial support of research activities, first through private philanthropies, then through government agencies, has given rise to the genre of the grant application. Here the researcher is expected both to summarize the present state of her research on the subject for which she requests support, and to outline her plans for future research. Although the historian must always evaluate grant applications with the caution that the applicant may present her plans in the light most favorable to the kind of research the agency prefers to support, these documents can still provide a precious type of information not ordinarily contained in more traditional documents: that is, the investigator's foresight, his anticipation of where the research trail still before him may lead. Comparisons of such predictions with the actual outcome of the investigative pathway subsequently followed during the grant period can greatly illuminate the manner in which scientists are constrained to adjust their plans and to change their courses in the face of unexpected results. Grant applications can give special force to the eloquent testimony of François Jacob that "if what we are going to find out is truly new, then by definition we cannot know in advance. There is no way to say where a given field of research will lead."[11]

A further type of documentation generated by the increasingly bureaucratic nature of financing for science in the last half century is the "trip report." To claim reimbursement for expenses, scientists must often state what value for them resulted from their travels to attend meetings, perform experiments in other laboratories, or consult with colleagues. Because much of the exchange of information that drives research forward takes place informally in such settings, even summary accounts of what was discussed can provide important clues for historical reconstructions.

In his study of experimentation in the "modern physics laboratory" of post–World War II particle physics, Peter Galison discovered that the types of documentation discussed above did not exist for the new "massive experiment" that had come to characterize this field. The physicists engaged in these large-scale collaborative ventures did not correspond with one another, nor did they keep meticulous laboratory notebooks as the physicists of earlier generations had. This dearth of traditional traces of their activity was, however, compensated for by the presence of new kinds of documentation necessary to sustain coordinated activity within such large groups. Among the many types of such records he uncovered, Galison found the most useful to be minutes of meetings, memoranda containing proposals and counterproposals, and claims and refutations circulated among the collaborators as data began to accumulate. From such new forms of "scientific literature," Galison found, he could piece together a record of the progress of such scientific teamwork nearly as complete as the older forms of record that track the progress of individual investigators.[12] Whether this shift in the nature of scientific records extends to other areas of recent science, or is peculiar to the dependence of experimentation in physics on huge, centralized research installations, remains for future historians to ascertain.

The individuals to whom I shall most frequently refer in order to illustrate more general features of investigative pathways are those I have myself studied. I shall here mention them in the chronological order of their historical appearance, rather than the order in which I encountered their lives and work. For the two earliest examples I have not yet completed or published detailed reconstructions, but introduce them here because their careers illustrate the probability that sustained research trails emerged as dominant patterns in the lives of early modern natural philosophers along with institutional structures that made them responsible to report regularly on what they had done to advance their particular sciences. Claude Bourdelin was among the first generation of chemists to be appointed members of the Académie Royale des Sciences in Paris at its founding in 1666. In its early years the Academicians were expected to participate in collective projects in which their individual contributions were not picked out for personal recognition. After several years of debate they undertook a project to collect plants from all over the world, identify and describe them botanically, and subject the matter of each plant to chemical analysis in the optimistic hope that they could discern differences in their composition that would explain their specific medical virtues. For most of the thirty-year period in which the project was continued, Bourdelin performed the chemical operations and recorded each result in a series of bound notebooks.[13] It is possible, therefore, to follow the day-by-day progress of this early example of organized, sponsored research, and to compare the ongoing investigative pathway with the frequent reports of its results that Bourdelin presented at weekly meetings of the Academy. Taken as a whole, the pathway is monotonously uniform. A complete reconstruction would be unrewarding. In the beginning, however, the project was imaginatively conceived, and there are, in its early phases, interesting shifts in method or emphasis, brought about by the intimate interplay between expectations and outcomes that did not fully sustain those expectations. The experimental trajectory traceable through these notebooks can be viewed as a primordial version of the modern research trail.

Etienne-François Geoffroy was one of the younger members of the second generation of Academicians. By the time he joined their ranks in 1699, members were expected to carry out individual research projects on which they regularly presented papers to their assembled confreres. The papers were then printed in the *Mémoires* of the Academy, which appeared annually (although the actual date of publication soon fell several years behind the nominal date). In the first decades of the eighteenth century most members of the Academy published regularly in this manner. Out of this institutional arrangement emerged the genre of the research paper as a kind of progress report on the latest research of the author, rather than as a broader survey of a field or problem. In a series of papers published by Geoffroy between 1702 and 1720, therefore, we can track, at a medium level of resolution, a representative investigative pathway of this period.[14] From his letters to Hans Sloane, the President of the Royal Society,[15] we can obtain a few glimpses behind the scenes of Geoffroy's published reports, but there are no surviving laboratory records to support a finer-structure analysis. Nevertheless, the density of his publications during this period is sufficient to outline some of the characteristics of investigative pathways that can be more fully characterized for cases in which such records do exist.

Antoine Lavoisier ranks among the foremost figures in the history of the early modern sciences. Through the investigation of the processes that "fix or release air" on which he embarked around 1773, and which he sustained in spite of many other activities and responsibilities until around 1790, when events of the French Revolution gradually overwhelmed him, Lavoisier introduced a new, thoroughly quantitative style of experimentation and reasoning into chemistry. The pursuit of these methods led him to a new theory of combustion, a new basis for determining chemical composition, a pragmatic definition of chemical elements, the first elementary analyses of organic compounds, the first specification of a biologically significant process (alcoholic fermentation) as a balanced chemical equation, and much else. By some he is regarded as the founder of modern chemistry, by others as the leader of a major reform in a science that had already established some of the scientific foundations that Lavoisier was able to merge with his own advances.

At the time he began systematically to pursue this ambitious and remarkably successful investigative pathway, Lavoisier also began to record his experiments in bound notebooks. Twelve volumes of these notebooks have survived,[16] and although there are some gaps, his research trail can be reconstructed confidently on the assumption that the notebooks contain most of the experiments he performed between February 1773 and October 1788. This record shows that Lavoisier was remarkably consistent in carrying out over nearly twenty years the agenda he had set for himself at the beginning. There are many short-term changes in the direction of his pathway, as he responded to local opportunities, temporarily suspended ventures that seemed not to advance, or was diverted by other occupations, but always his moves were connected with both his original objectives and the places to which his preceding path had led him.

Justus Liebig, one of the dominant chemists between 1830 and 1840, produced personally and with the numerous students that he trained in his laboratory at the University of Giessen a stream of publications on the composition and reactions of organic chemistry that played an important role in the emergence of organic chemistry as a major sub-field of chemistry. These papers appeared frequently enough so that it is possible to map Liebig's research trajectory in considerable detail in the absence of laboratory notebooks. In addition, his correspondence with Friedrich Wöhler provides glimpses of earlier stages in the research that led to some of his papers. For the several collaborative ventures that Liebig and Wöhler carried out during the 1830s, the letters reveal the fine structure of their experimental progress. Like most German professors in this era, Liebig had heavy teaching responsibilities and tended to crowd his own research into periods of intense laboratory activity during holidays and between semesters. It would be interesting to examine more fully than I have so far been able to do whether Liebig typically picked up where he had previously left off, or whether such schedule constraints also lent to his investigative pathway a qualitatively more intermittent character than those of scientists who are in a position to work continuously on their personal projects.

Claude Bernard, one of the most successful experimental physiologists of the nineteenth century, made a series of major discoveries between 1848 and 1860 that dazzled his contemporaries, and that have led some historians to elevate him to a rather mythical position as the founder of modern experimental physiology. In fact, he participated in a broader movement, taking place most prominently in France and Germany, which consolidated a field of investigative activity that had deep roots in earlier eras. Bernard left behind numerous laboratory notebooks, as well as many informal memoranda, notes to himself, and successive drafts of intended publications that enable one to retrace in great detail and depth the evolution of both his experimental work and the broader generalizations about experimental methods and biological organization that preoccupied him in his later years.[17]

Only selected portions of Bernard's laboratory records have been examined historically. They reveal that from the beginning he was a "diversifier" who maintained more than one special area of research. During his first ten years, he divided his work mainly between the chemical events of nutrition and the functions of selected portions of the peripheral nervous system, including in particular the sympathetic nerves. To these core concerns he added also the study of the effects of poisons such as curare and carbon monoxide. He moved frequently back and forth between these distinct pathways, the experiments of which he recorded in separate notebooks. Each pathway that he took up, however, he then pursued throughout his long scientific career. Even after 1860, when illness slowed the relentless pace he had kept up until then, Bernard returned again and again to the same set of problems that had engaged him in his early years.

Hans Krebs was the leading architect of intermediary metabolism during the decades from 1930 to 1950, during which the main reaction pathways through which foodstuffs are broken down to release energy, and essential molecules are composed through step-by-step syntheses, were worked out. His own two most important discoveries, that of the ornithine cycle of urea synthesis in 1932, and the citric acid, or "Krebs" cycle in 1937, provided the models on which the general conception of metabolic cycles was built. During the 1950s Krebs was also one of the first to turn from the problem of filling in the steps of these pathways to the problem of how such pathways are regulated. He maintained a prominent position within a maturing field until his death in 1981.

Krebs saved an unbroken series of laboratory notebooks from the time he entered the laboratory of Otto Warburg in 1927 until the outbreak of World War II.[18] In the postwar years, as he gave up his own bench work but continued to supervise closely the work of students, visitors, and technicians, the research carried on in his Metabolic Research Laboratory became more diverse and can no longer be traced through his own research records. The reconstruction of Krebs's trajectory for the decade and a half in which he worked alone or with one or two assistants reveals a representative, continuous investigative pathway. Beginning with the simple idea that he would apply the powerful manometric, tissue slice methods he had learned in Warburg's laboratory to elucidate metabolic pathways, Krebs shifted often during the early years from one pathway to another. When he sensed himself blocked in one direction he quickly moved in another; but all these oscillations took place within a set of problems approachable by similar methods, and with the common intent to identify intermediates in a chain of reactions interpretable by the known mechanism of organic chemistry. In later decades Krebs adopted new methods that entered the field, but applied them again and again to study at deeper levels the same set of problems he had taken up in the first decade of his independent research career.

Matthew Meselson and Franklin Stahl performed, in 1957, an experiment that provided strongly persuasive evidence that DNA replicates semi-conservatively in the manner predicted by the model of the double helix proposed four years earlier by James Watson and Francis Crick. The experiment was quickly hailed as a classic and has since been called "the most beautiful experiment in biology." Meselson had the initial idea from which the experiment eventually evolved, in 1954, at Caltech, where he was a student of Linus Pauling. Shortly thereafter, he met Stahl at Woods Hole, and the two agreed that they would work together on the project after Stahl came to Caltech as a postdoctoral student. Their short but intense collaborative investigative pathway began in September 1956, and took many twists and turns along the way to the first successful experiment in October 1957. Although conventional laboratory notebooks of their experiments do not exist, the preservation of the logs of the runs of the Model E analytical ultracentrifuge on which the experiments were performed, and of the ultraviolet absorption films constituting the immediate experimental results, permit the fine structure of their research to be reconstituted as a nearly continuous pathway.

During the 1950s Seymour Benzer carried out a project to map the fine structure of a portion of the genome of the bacteriophage T4 known as the "rII" region. The degree of resolution that Benzer reached approximated the dimensions of individual DNA nucleotides, and provided evidence compelling to his contemporaries that genes are sequences of DNA. Benzer's work is seen as the most visible site of the intersection between classical genetics and the emerging field of molecular biology. I am presently at an early stage in the reconstruction of Benzer's investigative pathway during the years 1954–1961, when he concentrated on this problem. The materials that Benzer has saved, including complete laboratory research records in the form of loose-leaf notebooks, extensive correspondence with other members of the early group of molecular biologists, grant applications, trip reports, and notes taken during courses and meetings he attended, constitute the fullest record of a scientific life during its formative period that I have so far encountered.

Benzer would appear, from a distance, to have broken the general pattern of the investigative pathway outlined in the introduction to this book. Trained as a solid-state physicist, he switched abruptly into phage biology in 1948. In 1965 he again switched from the field of phage genetics in which he had made his mark, to the behavioral genetics of the classical fruit fly. Closer examination will reveal, however, I believe, that even when making such leaps from field to field, Benzer found transition pathways that enabled him to cross from one main line to another without losing all the investments that he had made in his previous research activity. The first problems he took up in phage biology involved measuring the effects of ultraviolet radiation on the survival of bacteriophage in infected bacteria in order to follow the course of their reproduction, a genre of experimentation that had been initiated by other converted physicists, and for which he believed his own prior experience was relevant. Only after several years in this field did Benzer turn to a problem of genetic mapping for which that experience was much less pertinent. By then he had acquired sufficient skill and experience in his new field to do first-class work unrelated to his earlier work in physics.

This sampling of investigative pathways spans three and a half centuries. The conditions under which experimental science is pursued have changed drastically over that time. Seventeenth-century chemists operated with a small repertoire of traditional apparatus and operations, and little of what they then believed about the composition of matter has survived. Lavoisier introduced the practice of designing special apparatus and instruments for the new types of experiment he performed, beginning the process through which laboratories have ever since grown in size, complexity, and expense. Similarly, the theoretical complexity of the sciences pursued by these individuals has increased immensely over these same centuries. Has not the nature of experimental investigation become so different that there is little in common between the simple and repetitive analysis of a Bourdelin in the seventeenth century and the deeply imaginative, technically demanding experiments of a Meselson and Stahl or a Seymour Benzer in the mid-twentieth century? If, as Lorraine Daston maintains, the forms of rationality itself have evolved historically, the nature of what are recognized as facts and evidence changed during the seventeenth and eighteenth centuries, and the ideal of objectivity itself emerged only in the nineteenth century, how can we expect our subjects from the early modern era to have pursued their activities in ways comparable to those who have worked in much more recent or contemporary times?[19]

My own experience reconstructing investigative pathways over these centuries suggests that underlying the great changes in the scale and complexity of the enterprise, and even of subtler changes in the modes of reasoning or emotions associated with the activity, something fundamental has remained unchanged, and that it links those who have engaged in serious, extended enquiries into some aspect of the natural world through this entire period. The process by which natural philosophers or scientists identify problems that appear capable of solution with the means at hand, the interplay between questions asked and the responses of successive experimental outcomes, the mutually beneficial exchange of views between the individual investigator and his contemporaries engaged in similar ventures, the necessity for building on what has come before, even while subjecting previous solutions to critical reassessment, the strong tendency to persist in the pursuit of the problems to which one commits oneself early in a career, or else to move to areas to which one can bring one's previous experience—all these characteristics seem to have been shared by successful investigators over the last three centuries: that is, for the period in which science has existed as a continuous, organized, collective enterprise. In the following chapters I discuss in greater detail some of the patterns I have been able to identify in the work of these individuals that appear generalizable beyond their particular case histories.

Part Two **Phases in the Lifetime Career**

Chapter 3 Apprenticeship and the Achievement of Independence

Modern science may appear superficially to have little in common with the traditional craft guilds in which young men were required to serve a specified period of apprenticeship with a master before being permitted to begin such activity on their own. For nearly two centuries scientists have been educated in their fields during a long, academic process that begins with introductory general science courses in primary or secondary school, and introductions to the specific major sciences such as biology, chemistry, and physics in high school or university, to which the student returns at increasingly advanced levels in further courses. In the process future research scientists will have encountered many different teachers and will have learned much on their own by reading textbooks, solving problems, and carrying out exercises in teaching laboratories. Nevertheless, between the time in which they study science in general, and a special branch of science in particular, and the time in which they emerge as independent investigators, most scientists have undergone an extended phase in which they learn by doing, under the guidance of one or more experienced researchers in the field they expect to pursue.

One of the maxims often heard among twentieth-century scientists is that the most effective way to win a Nobel Prize is to be trained by a Nobel Prize winner. Genealogical trees are sometimes exhibited to show sequences of several generations of Nobel Laureates. Hans Krebs pointed out in 1967 that in his own "genealogy," he was the fourth successive generation of Nobel Prize winners trained by a scientist who himself had won the prize, and that the lineage extended back through four more generations of scientists who undoubtedly would have been winners if the prize had been in existence before 1900.[1] Statistically, however, a compelling case has not been made that this has been the privileged route to the now coveted honor of a royally sponsored trip to Stockholm. A deeper message underlying this folk-wisdom is that there are major benefits to be obtained by training with an extraordinarily successful older scientist: that by observing at close range how the master does it, the fledgling investigator can learn much about doing science at a high level that cannot easily be learned in other ways.

We can find many examples of scientists who achieved eminence on their own after working for some time with mentors who were themselves outstanding scientists. Among conspicuous nineteenth-century examples are Michael Faraday under Humphry Davy, Theodor Schwann under Johannes Müller, Justus Liebig under Joseph Gay-Lussac, August Wilhelm Hofmann under Justus Liebig, and Heinrich Hertz under Hermann von Helmholtz. The subsequent careers of two of the scientists I have studied in detail were powerfully shaped by such experiences.

Nearing the end of his medical training in 1839, Claude Bernard came under the direct influence of François Magendie, when Magendie offered him the position of *préparateur* for the course in experimental medicine that Magendie taught at the Collège de France. Accustomed to demonstrate before his classes experiments on animals that had previously led him to important discoveries such as the functional distinction between the dorsal and ventral nerve roots, Magendie required an assistant able to prepare the animals before each lecture, and perhaps to perform preliminary surgical operations on them. Whether Bernard had previously had opportunities to develop and to give evidence of such skills is uncertain, but Magendie must quickly have seen that the young future doctor had considerable operative talent, and have been a major factor persuading Bernard to aspire to a career in experimental physiology rather than in medical practice.

At the time of their encounter, Magendie was the most prominent physiologist in France, and one of the most influential in all Europe. He had begun in the first decade of the century as one among six or seven talented young graduates of a system of medical education reformed during the period of the French Revolution, whose training had included considerable surgical experience. Each of them had made substantial contributions to an emerging tradition of experimental physiology based on vivisection, but all except Magendie had either gone on to medical careers or died young, and he had become the primary spokesman and advocate for the importance of experimental physiology both as an independent discipline and as an essential foundation for medicine.[2] A forceful, persistent, and sometimes abrasive personality, Magendie must have been a formidable presence in Bernard's still relatively unformed life.

When Bernard began conducting experimental investigations of his own in 1842, the starting point for each of the lines of research he developed was a subproblem falling within a domain already pursued by Magendie. Some of the questions Bernard took up originated in experiments he had performed together with Magendie, the further steps of which he carried out alone. Sometimes he adopted techniques that Magendie had devised for one purpose and applied them to a different problem. In his early studies of digestion, for example, Bernard frequently relied on reactions such as that of potassium prussiate with iron, which deposited Prussian blue where the two reagents came together, that Magendie had earlier used to study the process of absorption.[3] The general style of Bernard's early investigations—the primary reliance on sometimes extremely delicate surgical interventions, and secondary application of chemical methods sometimes in collaboration with a professional chemist—was hardly distinguishable from the experimental style of his mentor.

Hans Krebs had completed his medical training when he encountered his most powerful mentor, Otto Warburg. Hopeful that he could combine a clinical career with research in a field such as internal medicine, Krebs had spent a year taking a remedial course in chemistry for aspiring medical investigators when he had the opportunity to enter Warburg's laboratory in Dahlem as a paid research assistant. There he learned how to perform experiments on the respiration of isolated tissue slices, whose consumption of oxygen and formation of carbon dioxide could be measured with great precision by means of a micro manometer that Warburg had designed by modifying a similar apparatus used earlier by the English physiologists George Barcroft and John Scott Haldane. Warburg did not organize his laboratory as a teaching institution for aspiring young scientists. Most of those who worked in his small quarters were technicians lacking advanced formal education, whom he trained to carry out the experimental operations he himself had devised, and who continued for long periods to study sub-problems within his own investigative pathway that he assigned them. Krebs was treated similarly.

During the three and a half years that he remained in Warburg's laboratory Krebs was given, one after another, problems on which Warburg had already made some preliminary experiments, and which he was confident that the inexperienced younger man would be able to solve by the application of the experimental methods currently practiced in the laboratory. To this extent Warburg nurtured Krebs into the experience of successfully completed research projects. When, however, Krebs had an idea of his own for how these same methods might be applied to the study of intermediary metabolism, Warburg told him that the problem did not interest him, and that there was only enough space in his laboratory to work on the problems that Warburg himself had initiated. Krebs made no further effort to take up an independent problem until after he had left Warburg's laboratory.

Otto Warburg was not only the preeminent biochemist in Europe at the time Krebs joined him, but a personality of great force as well as great eccentricity. He expected everyone who worked in his laboratory to be present promptly at 8 A.M. each morning, including Saturday, to experiment all day long, and to use his evenings to analyze the data acquired during the day. He accepted no reasons for taking time out during working hours to deal with personal affairs. He expressed his opinions forcefully, including his disdain for the work of other scientists of which he did not approve. Those who worked for him accepted his views with little question. He could be warm and helpful when his workers ran into problems, but he seldom gave praise. Krebs was clearly awed by Warburg's powerful intellect, his rigorous experimental methods, his disciplined style of work, and the pungency of his opinions.

It is commonly believed that the importance of such mentorship lies in tacit knowledge. By watching the mentor in action the apprentice learns much that cannot be articulated. This aspect of the relationship is certainly verified in my two cases. While carrying out vivisection experiments with Magendie, or observing Magendie carry out demonstration experiments on animals Bernard had prepared for him, the student undoubtedly learned much by watching carefully just how Magendie wielded the scalpel. For the first few days Krebs spent in Warburg's laboratory, Warburg instructed Krebs to watch while he himself performed the basic experimental operations on which the current investigations in the laboratory relied. Then Krebs began trying to do them similarly. Through the entire period in which he remained in the laboratory, Krebs worked just on

the other side of the same bench on which Warburg performed his own experimental work, and so had ample opportunities to learn by watching and imi-
tating.[4] But that is only part of the story. Much of what these students learned
to emulate were attitudes of their mentors, powerfully articulated during the
daily laboratory life they shared. Bernard followed Magendie's example, not
only in matters of technique passed on through watching and doing for himself
what he had seen Magendie do, but also in such major priorities as the urgency
of promoting physiology as a discipline essentially based on experimentation.
Krebs learned from Warburg the precept that one should not be afraid to "attack the great unsolved problems of his time," and to find the solution by doing
many experiments without hesitating over whether any one experiment was
worth doing.[5] Of course, Krebs did not just absorb the precept by listening to
Warburg; by watching his mentor in action he saw how Warburg went about
identifying the great unsolved problems and devising the experiments needed
to solve them.

Krebs not only attributed his own "good fortune" in science to having had
"an outstanding teacher at the critical stage of my scientific career," but noted
that Warburg had similarly testified that the most important event in *his* early
career was that he was able to spend three years working in the laboratory of the
great biochemist Emil Fischer. He was likewise aware that Justus Liebig had
stated that the "course of my whole life was determined by the fact that Gay-
Lussac accepted me in his laboratory as a collaborator and pupil."[6]

These crucial advantages of apprenticeship with an outstanding and forceful
scientist, however, also entail a corresponding difficulty afterward in emerging
from the shadows of the great man's eminence to establish an independent
course. In answering his own question, "what . . . is it in particular that can be
learned from teachers of special distinction?" Krebs wrote: "Above all, what they
teach is a high standard of research. We measure everything, including ourselves,
by comparison: and in the absence of somebody with outstanding ability, there
is a risk that we easily come to believe that we are excellent and much better than
the next man. Mediocre people may appear big to themselves (and to others) if
they are surrounded by small circumstances. By the same token, big people feel
dwarfed in the company of giants, and this is a most useful feeling."[7] That feeling dwarfed is most useful was, however, a perspective that Krebs could achieve
only after he himself had succeeded auspiciously. At the time Warburg told him,
in 1929, that he must leave the laboratory by spring of the next year, Krebs concluded that his mentor did not consider him capable of original research, and
he did considerable "heart searching" before deciding that he would try to go

on and test whether he could prove the opinion he believed Warburg to hold about him to be wrong.

When he came to Paris, the temperament of Claude Bernard was, as Mirko Grmek has put it, more poetic than classical. He preferred theaters and art museums, the romantic and the marvelous, to the rigors of academic reasoning and composition. It was Magendie who instructed Bernard in the need for skepticism and doubt, to mistrust doctrines and respect only facts. "What a hard school" for Bernard this was, according to Grmek, "when one thinks of the disciple as one with the ebullient, imaginative spirit of the poet" and of a master whose first inclination was skepticism and a "horror of theories."[8]

Bernard's early dependence on Magendie was compounded by a lack of professional opportunities in the Paris of the 1840s. Not only did his experimental efforts meet more setbacks than successes during the first five years, but his efforts to establish himself as a private teacher failed. For several more years he was forced to resume a subordinate position as a substitute teacher for Magendie's lecture courses. Bernard emerged from Magendie's shadow only in 1848, when he was fortunate enough to make two discoveries of sufficient magnitude to establish his own reputation as a leading experimental physiologist. Years later he was able to see Magendie's limitations as well as his strengths. According to Bernard then, Magendie had, indeed, founded experimental physiology, but his sternly empirical attitude had prevented him from developing deeper interpretations of the results he obtained. The suppression of his early penchant for romanticism in the long run served Bernard well. It tempered his imagination with a critical control that became his great strength, but only after Magendie's death did this combination come fully into its own.[9]

Buchwald has found similar relations between Heinrich Hertz and his mentor Hermann von Helmholtz, one of the most eminent scientists of the nineteenth century. Coming to Berlin as an advanced physics student in 1878, Hertz began his career as an experimentalist in Helmholtz's laboratory. His first project was to examine a prize question that was of great interest to Helmholtz, for the answer would distinguish between certain crucial aspects of Helmholtz's electrodynamics and rival electrodynamic systems such as those of Weber and Maxwell. From that day, Helmholtz "began to mold Hertz." Helmholtz came in for a few minutes every day to see how Hertz was progressing, and gave frequent, if somewhat distant support and encouragement. Hertz deeply imbibed both Helmholtz's conceptual structure of electrodynamics and his style of designing experiments: not to measure known effects more accurately, but to create and detect new effects.[10]

Quickly becoming a very effective experimenter, Hertz spent three years as an assistant to Helmholtz. He thrived under Helmholtz's tutelage, but soon came to realize that in the competitive atmosphere of the physics of his time, he also needed achievements that would distinguish him from his mentor. Although Helmholtz never prevented Hertz from taking up his own projects, he did exert gentle pressure to induce his student to be attentive to problems that interested Helmholtz himself. In 1883 Hertz received an offer to go to Kiel as a *Privatdozent* in mathematical physics. He accepted, despite the lack both of a laboratory in Kiel and of enthusiasm for the move from Helmholtz. One of Hertz's motivations, according to Buchwald, was to place "physical and eventually intellectual distance . . . between himself and Berlin."[11] Gradually Hertz began to evaluate more critically the deeper implications of Helmholtz's electrodynamics, and increasingly to set his own priorities; but he remained within the broader orbit of Helmholtz's theoretical and experimental style, and the "deep intellectual bond between the master and the apprentice" was never broken. Not until 1888, when Hertz's "tremendous success" in propagating electromagnetic waves made him an independent "center of contagion in physics" did he fully escape the role of gifted apprentice to the great Helmholtz.[12]

Each of these three found a pathway along which to move from his initial investigations under the strong influence of his mentor to the position of independence he eventually attained without sacrificing the experience he had gained through his earlier association with the master. Bernard continued to apply the experimental techniques and to follow the set of general problems he had begun as Magendie's assistant. He distinguished himself from his mentor more by the interpretative breadth he developed through disciplining his imagination with the skepticism he had learned from Magendie than by diverging from the laboratory methods he had learned under Magendie's guidance. When Krebs began research on his own after leaving Warburg's laboratory, he brought with him the powerful analytical methods Warburg had taught him. He both avoided direct competition with his mentor and found a field in which to establish himself, by adapting these methods to a set of problems that had not interested Warburg. Hertz distanced himself from the "deeper reaches" of Helmholtz's electrodynamics and resisted his mentor's efforts to draw him back to the problems that were of central concern to Helmholtz himself, but continued to perform experiments in the style he had learned during his years as assistant to Helmholtz in Berlin.

Bernard, Krebs, and Hertz each achieved independence and success of a magnitude that eventually equaled that of their outstanding mentors, but only after a period of struggle. Their ability to emerge as both independent and highly original is a measure not only of the skills they had acquired as apprentices, but of personalities robust enough to overcome the frequently intimidating effect of a powerful mentor. For each such protégé of an extremely successful master who has gone on to similar success, we could probably find several others who did not: people who either found themselves so dwarfed that they believed they could never measure up to the standards they had seen, and who gave up; or people who continued to imitate their mentors to such an extent that they did not achieve separate distinction; or people who, like the sorcerer's apprentice, wielded the tools of their mentor without the same skill or finesse. We know less about these cases, because historians of science do not ordinarily write about unsuccessful scientists.[13]

The three examples just discussed represent one end of a broad spectrum. Each of the three mentors was among the most powerful scientists of his age, each combined eminent achievement with a forceful, dominant personality. In each case the association between master and apprentice was close and prolonged over several years. The impact on the younger man was in each case profound. Although each of them also absorbed something from other more senior scientists in his surroundings, none of these influences competed with that of the single mentor in shaping the initial orientation of his protégé. Often the influence of mentors is more diffuse, especially in cases in which the fledgling investigator undergoes, either successively or simultaneously, the influence of more than one older person. The career of Antoine Lavoisier is illustrative of such patterns.

When Lavoisier began, in 1763, to show a clear preference for science over the law for which he had been trained, he did not immediately specialize. In the words of the geologist Jean-Etienne Guettard, with whom he traveled extensively on mineralogical and mapping expeditions during the next several years, Lavoisier's "natural taste for the sciences leads him to want to know all of them before concentrating on one rather than another." Some Lavoisier scholars, including Henry Guerlac, believe that Guettard "exerted the greatest influence upon Lavoisier and focused the young man's attention upon geology and mineralogy, and since it was an indispensable ancillary science, upon chemistry." It may have been at Guettard's recommendation, as Guerlac suggests, that Lavoisier attended the lectures of the most popular teacher of chemistry in France, Guillaume-François Rouelle. Historians generally have assumed that Rouelle was Lavoisier's principal teacher in chemistry, even though it is not known just when, or for how long, Lavoisier went to these lectures.[14] Recently several historians have attributed a great influence on Lavoisier to the lectures of the experimental physicist Jean-Antoine Nollet, which he also took in sometime in the early 1760s. Arthur Donovan has argued, in fact, that Nollet so inspired Lavoisier with his approach to science that it became Lavoisier's lifelong aim to make chemistry more like physics.[15] Too few details are known about the interaction of Lavoisier with either Rouelle or Nollet, however, to substantiate these conjectures about their respective influences on the aspiring young scientist.

In a doctoral dissertation as yet unpublished, Louise Palmer has examined large numbers of unpublished documents concerning Lavoisier's activities during these years that previous historians have used only selectively. Palmer has shown that interpretations that attribute a dominant influence of any single individual on Lavoisier's later directions underestimate the extent and variety of his early scientific occupations. In the years in which Lavoisier accompanied Guettard in the field, he also pursued chemical experiments in a laboratory he had set up in his father's house. He carried on chemical experimentation whenever he was not in the field. The two activities were closely related, in that he concentrated his experiments on the chemical composition of gypsum at the same time that he was deeply interested in the mineralogical properties and the mineral deposits of that substance. Together with Guettard, and on his own, Lavoisier traveled thousands of miles between 1764 and 1768 exploring the mineralogical and other physical features of France. With the botanist Bernard de Jussieu, he sometimes also went on botanical collecting excursions. Lavoisier regarded himself in these years as broadly and passionately engaged in the pursuit of natural history.[16]

From the earliest set of his surviving laboratory notebooks, Palmer has shown that as Lavoisier pursued the chemical analysis of gypsum in the privacy of his home, he was at the same time teaching himself the basic methods of analysis necessary to carry out his task. He had read extensively in the previous literature on the subject, but had to learn by trial and error how to perform the operations described in them, because he had not acquired these laboratory skills in an apprenticeship association. Palmer shows that Lavoisier was profoundly influenced by the ideas about chemistry he learned from Rouelle, but he evidently did not learn the practice of chemistry by working under the guidance of Rouelle or any other experienced chemist. It took Lavoisier nearly two years to succeed with his analyses. On the basis of these hard-won results he presented in 1765 a paper on the analysis of gypsum at the Academy of Sciences that was impressive enough to be published in the memoirs of the Academy, an honor probably due more to the literary skill with which Lavoisier was able to claim that his project had a broad significance for chemistry and mineralogy than to his analytical achievement itself. The methods that he applied to the work were less original than he thought, in part because in his self-study he had overlooked a paper on gypsum published more than a decade earlier by the eminent German chemist Andreas Marggraff that covered much of the same ground.[17] Lavoisier had proven himself able to learn on his own how to do a competent piece of chemical analysis using the well-known methods of the time, but he had not acquired the broad experience in this flourishing domain of contemporary chemistry that an apprenticeship period with an established chemist might have afforded him.

The association with Guettard was clearly the most intimate and long-lasting of Lavoisier's early career. Over the course of his fieldwork he gradually gained the experience and confidence to interpret what he saw independently, and eventually to reach the radical view that there may have been successive geological epochs marked by the alternating rise and subsidence of the seas. Had he continued to concentrate his attention on the geological and mineralogical problems he had studied with Guettard, the role of the latter would undoubtedly have been seen as that of his dominant mentor. When he was elected to the Academy of Sciences in 1768 and became a member of the tax-collecting agency known as the Ferme Générale, however, Lavoisier abandoned his life as a traveling naturalist, and only occasionally returned much later in his career to his youthful interests in geology and mineralogy. During the next several years he took up a variety of rather disconnected projects as he searched for a way to make his mark in the intense atmosphere of the Academy. For this stage in his scientific development there were no readily identifiable teachers. He had acquired independence, but not yet a sustained direction for his scientific ambitions.

In more recent periods, in which the methods that must be learned in order to become a practitioner in a given field are numerous and complicated, the opportunities for reaching a place at the forefront of a recognized experimental science by improvising, without prior experience in a well-equipped laboratory under the guidance of an experienced practitioner in that field, have probably dwindled to near the vanishing point. The pattern of preparation for independent research by working under a single master, however, has not necessarily become the predominant one. Aspiring young physiologists in the later nineteenth century, for example, often spent time in two or more of the well-known research centers that had by then emerged in Europe. A representative example is Ivan Pavlov. Introduced to physiological experimentation as an undergraduate student around 1874 by working in the laboratory of Elie de Cyon (Ilya Fadeevich Tsion), the newly appointed professor of physiology at the University of St. Petersburg, who had himself studied in the laboratories of both Claude Bernard and Karl Ludwig, Pavlov obtained his first overview of the field from Cyon's lectures, and learned in Cyon's well-equipped laboratory to perform vivisection experiments using modern graphic recording apparatus. Cyon had achieved considerable early success as an experimentalist, but did not rise to the first rank, and he imparted in his lectures mainly the physiological viewpoint of Claude Bernard, leavened by recent developments in German physiology. Under Cyon's guidance, Pavlov began work on the nervous regulation of the digestive glands and of the circulation, the former of which remained the driving force of his investigative pathway for the next twenty years. Pavlov was enormously impressed both by Cyon's masterful presentation of complex physiological issues, and by his "truly artistic ability to perform experiments." An offer by Cyon for Pavlov to become his laboratory assistant while studying medicine was forestalled, however, when Cyon was forced to resign his chair following a student disturbance. After completing his medical degree in 1884, Pavlov spent two years in the laboratories of Rudolph Heidenhain in Breslau, and Karl Ludwig in Leipzig. He came to appreciate the rigor of Ludwig's quantitatively oriented experimentation, but to prefer Heidenhain's emphasis on the purposefulness of the organism. As his own career developed, Pavlov was able to develop an approach that drew from each of these apprenticeship experiences, but was not dominated by the views of any single mentor.[18]

The practice that became increasingly standard after World War II, of following a Ph.D. program with several years as a postdoctoral fellow, has made it even more common for a young investigator to have served two or more apprenticeships, learning through successive experiences the style of more than one mentor, and emerging more easily from the potentially overpowering influence of any one of them.

The early career of Seymour Benzer illustrates how a succession of experiences in laboratories of leading scientists can teach, inspire, and lead a young scientist to practice according to the highest standards in a field while liberating him from the dominance of any one individual. As a graduate student in physics at Purdue University during World War II, Benzer was trained under the head of the department, Karl Lark-Horowitz, a physicist of considerable stature. As a member of a group of younger physicists around Lark-Horowitz, Benzer participated in a successful research project on the semi-conducting properties of crystals of germanium. Benzer himself contributed significantly to this group effort, and his future in solid-state physics appeared auspicious. Inspired, however, by

Schrödinger's *What Is Life?* to apply his knowledge of physics to biology, Benzer sought points of entry into a field completely new to him, but to which he believed he could bring his experience in physics.

Consulting Salvador Luria, one of the three leaders of the incipient field of phage biology, in nearby Bloomington, Indiana, Benzer was advised to write to Max Delbrück, who was building a center for phage research at Caltech, and had also founded a summer school at Cold Spring Harbor, through which hopeful young future phage biologists could gain their first exposure to work in the field. Delbrück sent a copy of a set of lectures he had recently delivered, which Benzer found as stimulating as Schrödinger's little book.

Benzer took the Cold Spring Harbor course in the summer of 1948, but Delbrück was not there that year, and the principal stimulus Benzer received was the pleasure of performing the relatively simple and quick, but quantitatively precise experiments on bacteriophage that constituted the hallmark methods of the phage group. Not as his first choice, but as his only immediate opportunity to begin working in phage biology, he then accepted a fellowship to do research in the recently established Biology Division of the Oak Ridge National Laboratory in Tennessee, beginning in the fall of 1948. There he encountered no older scientists of sufficient influence to make a strong impact on his direction of research. More or less on his own, he took up a recently published experiment by Luria and Raymond Laterjet, made some technical improvements, and oriented his research around problems of phage replication that he believed the experiment could illuminate. It had already been his aim, however, to go to Caltech to work in Delbrück's phage group. This he was able to do beginning in the summer of 1949. Delbrück did not disappoint his expectations. A few months after his arrival Benzer wrote back to Purdue that Delbrück was the "most inspiring person to be with." This was not apparent immediately on first acquaintance. "Many people," he added, " have formed completely distorted opinions of him, but we have begun to understand and share the devotion which the phage people feel toward him." Benzer absorbed much of the ethos that Delbrück maintained in both work and play; but perhaps because of his earlier experience in physics and because he had already made a start at Oak Ridge on an investigative project that he continued at Caltech, he was not so fully assimilated into Delbrück's style as some others, such as Gunther Stent, were. Delbrück's opinions of his work mattered to him, but did not dominate his directions. Nevertheless, the attraction of the place to him was strong enough to induce him to extend his stay at Caltech for a second year.

Following this episode Benzer had an unexpected opportunity to go to Paris, at the invitation of André Lwoff, to work for a year at the Institut Pasteur. Accepting enthusiastically, he found the laboratory there inspiring in different ways from Caltech. More formal and more reticent, Lwoff did not dominate his group in the same charismatic way that Delbrück did, but he was equally eminent in the field. Moreover, he had gathered around him other outstanding young and magnetic scientists, including Jacques Monod, François Jacob, and Elie Wollman. In Paris Benzer experienced a different scientific style and a different approach to the study of bacteriophage, and he took up problems different from those he had previously been studying.

After these four peripatetic years, Benzer finally returned to Purdue in the fall of 1952. There he was the only phage biologist, a fact that at first made him feel isolated. Soon, however, he began to make frequent trips to Urbana, Illinois, for meetings held in Luria's phage group. He kept also in touch by correspondence with the other centers in which he had spent time. At Purdue Benzer began the project on mapping the fine structure of bacteriophage T4 that led him to his best-known contribution to phage biology. The project carried to a much finer degree of resolution a kind of mapping, transferred from classical fruit fly genetics to phage, that had been begun not in the laboratories in which Benzer had personally spent time, but in those of two other prominent phage biologists, Alfred Hershey at Cold Spring Harbor, and Gus Doermann at Rochester.

When he had obtained his first significant results in mapping the T4 genes, Benzer wrote a paper, in the summer of 1954, and sent it to Delbrück, who lauded the work but expressed skepticism about Benzer's conclusions. In further encounters, Delbrück continued to doubt that Benzer's results validated the Watson-Crick model, according to which genes were linear sequences of nucleotides. Although Benzer valued Delbrück's views, which he regarded as a salutary moderating influence on him, he did not give way on any of the main points of his argument.[19] Benzer had been inspired, but not intimidated by the forceful leader of the phage group. In part his self-confidence may have derived from an innate independence of character, but it was undoubtedly enhanced by the multiplicity of his experiences in the laboratories of highly successful older scientists. Inspired from the beginning by Delbrück from a distance, Benzer had already acquired some experience in phage biology by the time he encountered the famous phage leader at close range. His admiration for Delbrück was then tempered by further experience in a laboratory where investigators of at least equal eminence worked. Consequently, Benzer seemed to experience little difficulty emerging as a fully independent investigator. Not dominated by any single one of the multiple influences on him when he was finding his way into a new field, he emerged from these several apprenticeship experiences with a project that was as distinctive as it was favorably received in the entire phage community.

Matthew Meselson achieved, even while still a graduate student, an early independence from powerful mentors, in part because of the originality of his own idea for an important experiment, though in part because he came under the influence of more than one senior scientist. A student of Linus Pauling, one of the most dominant scientists and forceful personalities of the time, Meselson pursued a Ph.D. project assigned to him by Pauling, to establish by X-ray crystallography the structure of a molecule whose peptide bond resembled that of the bonds linking amino acids in proteins. This was a small piece in Pauling's larger effort to establish the three-dimensional structure of proteins. Meselson followed the procedures characteristic of Pauling's group and produced a result that fit unremarkably within the ongoing flow of research of the laboratory. It was a nice piece of work, but it in no way enabled Meselson to stand out as an X-ray crystallographer or structural chemist, and he never, in fact, published anything on the subject. Meselson himself felt that, although he was competent in this field, he did not grasp the methods of X-ray crystallography at a deep enough level to do original work.

Meselson discussed with Pauling his idea to use density difference methods to test current views about the manner in which DNA replicates, but Pauling advised him to finish his thesis first. Meselson followed this advice, and when he had obtained a structure for his molecule, he began, in collaboration with Franklin Stahl, a postdoctoral fellow in Delbrück's phage group, to explore means to implement his idea. As he did so, he came into increasing contact with Delbrück, who was very supportive of the project even though it did not fit within the ongoing research directions of his group. Meselson also encountered through the project Jerome Vinograd, an expert in centrifugation, who taught him how to use the analytical ultracentrifuge on which he and Stahl intended to perform the experiments. Pauling generally approved of Meselson's new activity, but remained in the background.

It took Meselson and Stahl a little more than a year of intense effort to produce the remarkable result now universally known as the Meselson-Stahl experiment. That they were encouraged to focus on an investigation unrelated to the activities of either Delbrück's or Pauling's laboratory and given the resources necessary to carry it out was due to a special ethos at Caltech that supported independent initiatives by graduate students and postdoctoral fellows. The leadership style of Delbrück was especially important. Though given to strong opinions and sometimes harsh criticisms, he also possessed a democratic spirit, a belief that good ideas could come equally from senior or junior scientists.

The great success of their venture gave Meselson, in particular, an early self-assurance, as well as rapid recognition as a young scientist of unusual talent and achievement, which enabled him from then on to chart an independent pathway into further investigative problems. Many of the problems he took up exploited the density gradient method that he had devised in the course of the DNA replication experiments. Both Delbrück and Pauling remained for him important role models, but in part because their own styles were very different, neither of them dominated his own style or choice of future research directions.

The examples I have given suggest two broad types of mentor-apprenticeship relations: the profound effect of a single, powerful mentor and the necessity afterward to establish a separate identity on the one hand, and the more diffuse effects of two or more mentors on the other. The personal experiences through which scientists have begun careers that have moved on to notable achievement do not, however, always fit into such patterns. The endless variety of individual personalities and talents has produced many variations on the nature of the interactions through which scientific investigation is passed from one generation to the next. The American geneticist Herman J. Muller entered the "fly room" at Columbia University, composed of Thomas Hunt Morgan and two other students, Alfred Sturtevant and Calvin Bridges, around 1910, just at the time when that group was beginning to discover the many mutations of the fruit fly, *Drosophila melanogaster,* on which they constructed during the next five years the field of classical genetics. Muller recognized, as he put it later, that Morgan's "evidence for crossing over and his suggestion that genes further apart cross over more frequently was a thunderclap, hardly second to the discovery of Mendelism." Frequenting the fly room during these years, Muller, who had already decided to become a geneticist, served his scientific apprenticeship learning to perform experiments on Drosophila in the style being developed there; and the techniques as well as the basic approach he learned became the foundation for his own later career. According to the recollection of Sturtevant, this "group worked as a unit. . . . There was little attention paid to priority or to the source of new interpretations. What mattered was to get ahead with the work. . . . There were many new ideas to be tested, and many new experimental techniques to be learned. There can have been few times and places in scientific laboratories with such an atmosphere of excitement and with such a record of sustained enthusiasm. This was due in large part to Morgan's own attitude, compounded of enthusiasm, combined with a strong critical sense, generosity, openness, and a remarkable sense of humor." Despite the heady nature of this experience, however, Muller was not awed by Morgan or by the group ethos. More theoretically oriented than the others, Muller quickly became critical of some of Morgan's explanations of his experimental results. He thought Morgan's aversion to theoretical analysis was limiting, and that Morgan and Sturtevant's interpretation of crossing over was too simple. Worse still, when Muller attempted to analyze such situations more deeply, he came to feel that the "group" absorbed his ideas without acknowledging their source. He believed that the "theory of the gene" must be established by more complex analyses than Morgan and the others had undertaken, and he pursued these ends, first within the group, and after he had left it, on his own. Whereas they were, he thought, content to view the gene as a "bead on a string," he made it his central objective to determine the properties of the gene that enabled it to reproduce its own mutations. Muller acquired tacit knowledge of Drosophila genetics and genetic mapping by doing what the other members of the group did in the fly room, but from the beginning he did not "fit in" temperamentally with the rest of the group, and he very early became intellectually independent of the others. Muller's subsequent struggle was to establish his professional independence, a goal much more difficult for him to attain.[20]

Methodologically very competent scientists who have not themselves made highly original discoveries are sometimes able to train and inspire more imaginative students who are subsequently able to do more original research with the methods learned than had those who taught them. Gerald Geison has described an illustrious example in the career of the English physiologist Michael Foster, founder in the late nineteenth century of the Cambridge School of physiology. Mainly an inspiring teacher who only briefly carried on original experimental work, Foster was able to attract and train a group of extremely able younger men, including Walter Gaskell and John Newport Langley, whose investigations of the autonomic nervous system became pathfinding.[21]

One can easily find still other patterns of mentor-apprenticeship relations. The cases described here, however, can illustrate for us the importance and the complexity of the process through which most individuals have had to pass to make the sometimes difficult transition from advanced student to independence—taking those first critical steps that mark the beginning of a new investigative pathway.

Chapter 4 Mastery of a Domain

In the older, heroic view of the origins of modern science the great scientists of the past were often portrayed as possessing superior insight and ability from the time they first entered the field. With unerring foresight of the road ahead, they required only a well-planned preparatory period to establish the solid grounds on which they would challenge existing theories or conventions. Typical of such accounts was the late nineteenth-century scientist-historian Marcellin Berthelot's description of Lavoisier in 1773 at the start of his investigation of the processes that absorb and release airs: "In the solitary meditations of his laboratory Lavoisier formed the project of an enterprise whose character and scope he perceived from the beginning, . . . and he achieved his enterprise with a method, a continuity, an invincible logic, while employing in due measure in the pursuit of his plan the facts already known and the particular discoveries that a group of men of genius, his contemporaries, were making every day, equally skilled experimentalists, perhaps more original than he in detail, but whose minds were less powerful."[1] Jean Louis Faure wrote in 1925 of Claude Bernard: "From the first years, the marvelous accumulation of his discoveries struck the scientific communities with astonishment. In spite of the somewhat disorganized efforts of a person like Magendie, physiology could be said not to have existed. Today we know it as a fully constituted science, thanks precisely to the work of Claude Bernard." Faure repeated a story according to which, on the third or fourth time that Bernard prepared the animals for Magendie's course demonstration experiments, the latter "walked out of the lecture hall, declaring in the churlish tone that was habitual with him, 'All right, you are stronger than me.'"[2]

Over the decades since, we have come to see great scientists as more ordinary humans, not necessarily more brilliant than others whose careers remained unexceptional. "Expert performance" is now seen by some as more a matter of continual practice and growth than of inborn genius. Even those whose exceptional talents warrant their classification as geniuses seldom know when they set out on their first independent investigative ventures the destinations to which their pathways may lead them.

In his study of seven great creative individuals of the "modern" era in fields ranging from physics to dance, Howard Gardner has concluded that "no matter how potent" the initial "intoxication" of the creative person with the domain of activity in which he or she later achieved eminence, at least ten years of steady work at a discipline or craft seem required before that métier has been mastered. The capacity to take a creative turn requires just such mastery, and accordingly, significant breakthroughs can rarely be documented before a decade of sustained activity has been accomplished. Even Mozart had been composing for at least a decade before he could regularly produce works that are considered worthy of inclusion in the repertory.[3] In science, such mastery begins, but is seldom completely achieved, during the period of apprenticeship described in the previous chapter.

When Antoine Lavoisier ended his only real apprenticeship, to the geologist Guettard, and turned away from the mineralogical and other field activities for which he had acquired the most solid preparation, he had only begun his struggle for mastery. Carl Perrin has noted that from the beginning of his interest in science, Lavoisier had been searching for a problem that would lead him to "une belle carrière d'experiences" ("a fine course of experiments") that would enable him to make a mark in one of the sciences that interested him. In other words, he was seeking an opening onto a productive investigative pathway. He had many such ideas, some of which he implemented, but none of them proved starting points for continuous research. After publishing two papers on gypsum, he went no further in the directions his initial studies pointed. Late in his fieldwork he began measuring the specific gravities of the mineral waters found in the regions in which he traveled, using a hydrometer of his own design. In 1768 he presented a paper touting his hydrometric method as superior to the traditional qualitative analyses of mineral waters, but he did not pursue the method to any significant outcome after giving up his fieldwork. He next performed the experiments to disprove the supposed transmutation of water into earth to which historians have given special attention because they seem to mark his early attachment to quantitative methods; but his success in this endeavor did not lead to anything beyond the refutation of an old idea that had come down from the seventeenth-century willow tree experiments of Van Helmont, which had seemed to substantiate Van Helmont's belief that all substances are composed of water.

His early experimental efforts illustrate both Lavoisier's extraordinary talent and the difficulties that even such a talent can encounter finding a place for himself without the guidance of a mentor within a well-defined domain of investigative activity. Finally, however, he came upon a problem that afforded him the opportunity for which he had been so long on the lookout. When he found in the fall of 1772 that phosphorus and sulfur gain weight if they are burned, and that a calx of lead loses weight and releases air when it is reduced, these partially accidental discoveries launched him on yet another fresh investigative venture. In February 1773, he set out an agenda for the investigation of processes that absorb or release air that he anticipated would lead him to results of great consequence for physics and chemistry. The processes that would engage him—respiration, combustion, calcination, and "certain chemical combinations"—were stated very generally. Nor did he begin with a clear knowledge or mastery of the methods he would have to use to examine these processes. During the next few weeks he tried with limited success, using combinations of apparatus adapted from conventional chemical laboratories and the "pneumatic" apparatus of Stephen Hales, to implement his program. Most of his early experiments either failed or produced indecisive results. What is most evident from a close examination of Lavoisier's investigative pathway over these first months was that he was improvising, making his methods up as he went along, learning from his many mistakes how to improve the design and operation of his apparatus. By the late summer of 1773 he had learned much better how to control such processes than he had known in the spring. By then also he had disproved his own early theory that in all these processes it was the "fixed air" discovered by Joseph Black that was absorbed or released.[4]

The contrast between Lavoisier's position as he embarked on this course and that of three of the scientists—Bernard, Hertz, and Krebs—discussed in the previous chapter is striking. They left positions in which they had spent time as apprentices with well-honed methods and skills acquired under the guidance of their mentors. Their most pressing need was to establish through their own work independent identities and achievements. Lavoisier began his new investigative effort with little prior experience in the type of experimentation needed to pursue his problem. The originality of his plan was clear from the beginning, but the naïveté of his first efforts is also evident. The skills necessary to carry it out he acquired through the hard lessons of his early failures.

We might at first be tempted to attribute these differences to the fact that Lavoisier entered his scientific career in an age in which the skills to be learned were much less developed than in the time of Bernard, Hertz, or Krebs; but that would be to underestimate the amount of both articulated and tacit knowledge that eighteenth-century chemists possessed. Lavoisier might well have got his program under way with fewer troubles if he had spent time in the laboratory of one of the leading chemists of the time.

If his inexperience slowed his early progress, it did not, in the end, limit the achievement of Lavoisier. During the six months in which he struggled to make his experimental systems work, he essentially invented a new method of chemical experimentation, in which he weighed or calculated the weights not only of the solids and liquids that chemists had traditionally weighed, but also of the "airs" that entered or left during a chemical operation. As he continued to improve these methods, he came to command them in a manner none of his contemporaries could match. Through his own resourcefulness, determination, and imagination, Lavoisier more than compensated for the lack of the kind of apprenticeship that has launched many other scientists onto careers of outstanding achievement.

Lavoisier's ability to lift himself past others without the benefit of having observed his science performed at the highest levels under the guidance of a more experienced chemist was, however, conditioned by the fact that he came to the field at a particular juncture in which novel methods proved necessary to exploit fully the recent discoveries of new species of "airs." Lavoisier never attained the level of skill in more conventional chemical analysis that enabled more thoroughly trained practitioners of that art, such as Torbern Bergmann and Carl Wilhem Scheele in Sweden, to make prolific discoveries of new acids and bases during the same years that Lavoisier was establishing the new methods appropriate to the relatively unexplored field of activity that he took up in 1773.

Mastery of his new methods was only the first step in Lavoisier's quest to master the field of investigation he had laid out for himself. After the demise of his first theory that fixed air is the matter exchanged in each of the various processes that concerned him, it took him more than three years of intense experimental effort and thought to replace it with the general theory of combustion he proposed late in 1777. Although some accounts of Lavoisier's progress during these years depict him seeing clearly already in 1773 where he was headed and, thereafter, only biding his time while he gathered sufficient evidence to attack the prevailing phlogiston theory publicly, a detailed reconstruction of his investigative pathway over these years shows him feeling his way, unable to integrate his shifting explanations for individual phenomena into a coherent theoretical structure, sometimes continuing to reason in terms of phlogiston, sometimes ready to reject it, often following experimental leads provided by his English counterpart Joseph Priestley, sometimes tempted to give up his own partially constructed theoretical framework in favor of Priestley's views. There is not space here to follow the twists and turns of Lavoisier's investigative pathway in detail,[5] but one can, in summary, say that it did require, if not a full decade, at least five years of prolonged, steady effort to master the craft and the domain of inquiry which he had in 1772 recognized as holding the potential for the "fine course of experiments" for which he had been searching.

Claude Bernard similarly required about half a decade of relentless effort fully to master the domains of experimental physiology through which his early investigative pathway ran. His basic skill in vivisection was probably exceptional from the beginning, although he gradually expanded the repertoire and difficulty of the operative procedures on which he relied. Most deficient in the first years of his work on the processes of nutrition was his knowledge of the chemical methods applied to identify chemical processes in the organism. He made up in part for the lack of an extended apprenticeship in chemistry by collaborating with the young chemist Charles Barreswil, but did not achieve a level of skill at chemical analysis and interpretation comparable to his prowess with the scalpel. Nor was he sufficiently familiar with the range of current and recent investigations by others, particularly in the German-speaking lands, on the problems he took up, to benefit fully from the advances they had made. Consequently, the theories of digestion that he proposed between 1843 and 1845 were both somewhat idiosyncratic and ephemeral. In the next two years he published some interesting and original observations, such as the manner in which the urine of fasting herbivores becomes like that of carnivores, but he made little progress in his larger quest for a general theory of gastric and intestinal digestion. Moreover, in contrast to his later insistence that one should discard a hypothesis whenever an experimental result contradicts its predictions, he sometimes clung stubbornly to his favored ideas about digestion in the face of repeated failures to confirm them experimentally. Looking back on his early struggles from the perspective of his later successes, Bernard wrote: "For more than two years, at the beginning of my career, I wasted my time pursuing theories and chimeras. . . . I insisted on repeating experiments which insisted on responding to me . . . contrary to my views. It was not until after a long deception that I ended by reflecting that the struggle was not equal and that my will could not change the laws of nature and that I could do no better than to follow the indications of natural phenomena, using theories as torches intended to illuminate the path and needing to be replaced as they were consumed."[6]

The detailed reconstruction of Bernard's investigative pathway during these years reveals no single point at which such an insight appeared to have altered his approach; rather he persisted in pursuing unsupportable theories until near the threshold of his first two "breakthroughs," in 1848, the discovery of the special action of pancreatic juice on fats, and the discovery that the liver secretes sugar into the blood. After these two successes Bernard seemed to make further discoveries with ease (Figure 1). Whether his successes derived from a change in attitude, or

Claude Bernard in 1849. From J. M. D. Olmsted collection. Photo courtesy of Ralph Kellogg.

the latter was made possible by his successes, one could say that with these two discoveries Bernard had finally achieved the hard-won mastery over his métier that some historians have portrayed him as having possessed from the start.[7]

The pathway to first auspicious success was for Hans Krebs much shorter. After leaving Warburg's laboratory, Krebs spent one year at a municipal hospital near Hamburg. There he had opportunities, in the midst of clinical responsibilities, to carry out some research. In return for Warburg's support for a grant to purchase manometers, he was constrained to continue investigating problems of interest to Warburg that Krebs himself regarded as too routine to be interesting. He found time also, however, to take up his first independent project, a follow-up on the recent discovery by Einar Lundsgaard that iodoacetate can block glycolysis in animal tissues without inhibiting their respiration. Using yeast and slices of various animal tissues in the Warburg system, Krebs was able to extend Lundsgaard's result and draw some tentative conclusions about the relation between glycolysis and the aerobic phases of carbohydrate metabolism. The work resulted in the first publication based on his independent investigative venture (he had already published 28 papers derived from the projects Warburg had assigned him). It was not a major achievement, but it did mark a significant first step in the implementation of the general idea he had had as an assistant to Warburg, to apply Warburg's methods to the elucidation of intermediary metabolism.[8]

The following year, having moved to Freiburg, Krebs had somewhat more ample opportunities for research, though he was still busy as a clinician. There, after several months, he took up a new application of Warburg's methods to intermediary metabolism, studying the synthesis of urea in animal tissues. The outlines of the reaction—that two molecules of ammonia derived from the amine groups of amino acids combine with carbon dioxide to produce urea— had long been known, but a succession of efforts to determine the immediate steps through older methods had been unsuccessful. For several months Krebs and his assistant, the medical student Kurt Henseleit, were able only to confirm the known reactions in rat liver tissue slices. An unexpected result—that the addition of a rare amino acid named ornithine together with ammonia greatly increased the rate of formation of urea—however, directed Krebs's investigative pathway in a new and promising direction. It required him several months to elucidate further the nature of the ornithine effect and to explain it. When he had done so, proposing a cyclic process in which ornithine acts as a catalyst, Krebs had made a major discovery, one that brought him immediate attention within the biochemical community (Figure 2).[9]

Hans Krebs in 1932. Photo courtesy of Hans
Krebs.

That Krebs was able to reach such a place on the forefront of a major sub-field
of biochemistry just two years after becoming independent of his apprenticeship is impressive, and in part due to the diligence and resourcefulness that he
displayed as soon as he was given the chance to follow his own research pathway. He was, however, able to bring to the task powerful methods inherited from
Warburg, in the use of which he had already acquired four years of experience.
He had, therefore, already mastered the central skills necessary for success in his
new arena.

This auspicious initial success did not necessarily mean, however, that Krebs
had so quickly reached full mastery of his field of investigation. Although the
manometric tissue slice methods he had learned from Warburg provided the basic methodological foundations for his research, he had to supplement them by
devising micro-analytical methods for identifying the intermediate compounds
he thought might be produced in the various metabolic processes to which he
next turned. Gradually, during the next few years, after he had come as a refugee
to the Biochemistry Department in Cambridge, he expanded the repertoire of

such analytical methods available to him. In his last year at Freiburg and his first at Cambridge, his efforts to elucidate the general stages of the breakdown of carbohydrates and fats led nowhere. During the years that culminated in his most important discovery, the citric acid cycle, in 1937, one can observe a steady maturation of Krebs's investigative style as his accumulating experience led to ever fuller mastery of the domain he had chosen to make his own.

The period of apprenticeship and that of progressive mastery as an independent investigator were, for Krebs, sharply demarcated. The former took place in Warburg's laboratory, the latter after he had left it. For Seymour Benzer these phases of his career as a phage biologist were more nearly coextensive in time. Only in the six-week summer school course that he took in 1948 did Benzer learn to perform the most basic experiments on bacteriophage as a series of student exercises. Because he had shifted into this field after obtaining his Ph.D. in physics, Benzer learned his new field as a postdoctoral fellow, and he began immediately after this brief training experience to carry out his own investigative projects. He went, therefore, to Caltech to continue along an investigative pathway he had identified for himself at Oak Ridge, even though his purpose in doing so was to benefit from the mentoring of Max Delbrück and other more senior investigators working in Delbrück's group. Similarly, when he went to Paris he devised his own research project but learned from Jacob, Wollman, and others about lysogeny and other aspects of phage biology that had been developed there.

The outcome of Benzer's hybrid trajectory as part apprentice, part independent investigator was mixed. That he was able within a few months at Oak Ridge to teach himself to perform phage experiments with considerable competence was due, probably, in part to his own talents and discipline, but in part also to the fact that the phage biologists relied on a small repertoire of relatively simple experimental techniques that they repeated with endless variations to address different questions, and in part to the fact that he chose for his first problem the effects of radiation on the replication of phage, a problem to which he could bring his prior experience in physics. The fact that as a graduate student in physics he had learned general strategies of experimental investigation transferable to other domains probably also accelerated his progress in self-learning.

As his first project, Benzer fixed on an experiment recently published by one of the leaders of phage biology, Salvador Luria, together with Raymond Laterjet. By irradiating bacteria infected with phage with several doses of ultraviolet light at several intervals during the "latent" period between the time of infection and the time that the bacteria lysed, releasing progeny phage, Luria and

Laterjet had hoped to learn something about the nature of the phage replication process, in particular about the number of phage particles present during successive phases of this process. The curves that they obtained, however, did not confirm the predictions of "target theory," and they were unable to give their results an interpretation that illuminated the nature of the replicative process during what was known as the "dark" period. Benzer made several modifications in the procedures Luria and Laterjet had used, enabling him within a few months to be able to perform the experiment with greater precision than had its inventors. At Oak Ridge and Caltech Benzer continued to improve the experiment and to obtain with it impressive curves showing an increase in the resistance of the phage to radiation over the course of the latent period.[10]

Delbrück had welcomed Benzer's proposal to bring this project with him to Caltech, but had warned him, "You may get beautiful curves from these experiments, but I doubt that the curves will be interpretable in as simple a manner as you hope." The prediction of the more experienced phage biologist proved accurate. Just before Benzer left Caltech for Paris, Delbrück pressed him to write a paper on the results of his experiments. In that paper the most that Benzer could claim for the three years he had spent on the work was that he had obtained a tool that "offers promise . . . for studying phage growth." It was, he acknowledged to his former physics mentor, Lark-Horowitz, only a progress report. Benzer had already become a virtuoso experimentalist in phage biology, but he had not yet produced anything to advance significantly the understanding of the replication process that his experimental system was designed to study. He left Caltech in some sense still an apprentice in his new field. In Paris, too, he produced a paper that impressed such leaders of the field as Alfred Hershey with the high quality of its reasoning and results, but not one that made a landmark advance. During his four years away from Purdue, however, Benzer had progressed from novice in a new field to one who fully mastered its methods, its ethos, and its aims.[11]

Back finally at Purdue, Benzer came across early in 1954 the opening that moved him from mastery to distinctive achievement (Figure 3). Through a series of partially accidental coincidences, he recognized that he had at hand an experimental system that would enable him to map bacteriophage mutants with an unprecedented degree of resolution. Turning his accumulated experience in what was for him a new investigative direction, he was able within a few months to produce the first promising genetic maps of what was designated as the rII region of the bacteriophage T4 genome. Further refining his techniques, he reached in the following year a level of resolution approaching the dimensions

Seymour Benzer at Purdue University during the 1950s. Photo courtesy of S. Benzer.

of several DNA nucleotides. By 1957, his work was seen to have provided the bridge connecting classical genetics with the newly emerging molecular biology.[12]

Of all of the examples we have been considering, Matthew Meselson and Franklin Stahl moved most rapidly through the stages of mastery to landmark achievement. Setting out in their collaborative venture in September 1956, they followed a succession of changes in direction of their research pathway that sometimes diverted them from their original goals, but they were always able to return to their first intentions, and slightly more than one year later performed the experiment that made them famous. How were they able to move into a new arena and arrive at such a result so quickly? In the first place, their idea was of such compelling originality that they were not competing with anyone more experienced with similar intentions. Second, the accidental discovery that CsCl formed a density gradient in the centrifuge cell provided them with a novel method. Even though a novice at the operation of the analytical ultracentrifuge, Meselson was soon using it in a way that no one before him had done. Not only did the new method enable him and Stahl to succeed in their particular investigative venture to test the modes of replication of DNA; Meselson was also able to turn the method to other uses, and other, more senior scientists also turned to him, as the reigning expert, to learn how to apply it to their problems.[13]

According to her most recent biographer, Nathaniel Comfort, Barbara McClintock possessed such "dazzling cytological skills" from the time she first peered through a microscope as a graduate student at Cornell, that she almost immediately outshone her teachers and her fellow students in the identification and analysis of chromosomes in maize plants. Her ability to distinguish each chromosome at the pachytene stage gave her a basis for a series of early papers in maize genetics, six of which appeared in major journals within two years after she obtained her Ph.D. degree in 1927. Forming the nucleus of a small group of cytogeneticists, she quickly established a reputation for publishing brilliant, though difficult papers.[14]

Was McClintock an exception to the view that great scientists require an extended period of steady work at a discipline fully to master it? Did her extraordinary visual gifts, and particularly her ability to recognize complex patterns, enable her to pursue with unerring foresight the independent pathway she followed? Comfort's account of McClintock's long experimental pathway suggests that her special skills did enable her quickly to excel her colleagues in maize genetics, but that to construct a "developmental genetics of her own" nevertheless required two decades of relentless experimental work, and that she only gradually acquired the insights on which she based the ideas about the transposition of genes and the control of mutations and development that made her viewpoint so distinctive.[15]

My examples suggest that, in science as in the arts, steady work at the discipline or craft is required to master a domain, and that only after such mastery has been attained is the scientist able to produce results that advance the field in which he has learned to practice at a high level. They suggest also, however, that the "ten-year" rule is too rigid to apply without modification; that the period required may be nearly that long, but that it can also be much shorter, depending on particular circumstances. Intuitively we might expect that the period of preparation would be longer in the sciences than in the arts, because of the degree of rigor and specialization that must be assimilated to reach the levels of performance of those already established in the discipline. We tend to think of artists as more spontaneously creative than scientists, able to express true genius and originality more quickly. The opposite, in fact, may be true. Although painters, composers, sculptors, musicians, actors, or dancers develop individual personal styles, in general they must compete with all others in the same overall artistic genre to attain both the expertise and originality to gain eminence. The very high degree of specialization within science that can, on the one hand, demand prolonged periods of training to master its deeper conceptual structures and complex investigative methods, may also open niches in which the young investigator can more rapidly reach the forefront. When new methods are needed, or found before the need is noticed, as in the cases of Lavoisier and Meselson and Stahl, those who invent or discover such methods need not compete with more experienced investigators in their use, because there are none. On the other hand, where the methods are more obvious and have been long in use, then a longer period is required for newcomers to catch up to the level at which their mentors and other older investigators have already been operating.

Whether the period is long or short, however, our examples testify to the fact that even extraordinarily talented young investigators do not arrive with clear knowledge of the road ahead. With or without deep involvement with a single mentor, they, too, must take the first tentative steps along their investigative pathways without seeing in advance just where the trail will lead. They must sometimes stumble along the way before they gain the sure footing that will enable them to stride with confidence toward the achievements that may mark them as major discoverers and leaders in their fields.

Chapter 5 Is Distinction
Achieved or Conferred?

In the days when historians of science confidently assumed the permanence and reality of the progression of discoveries that collectively make up the structure of the modern sciences, they credited the scientists of the past with having made such discoveries at whatever point in their investigative trajectories surviving documents attest that they had first recognized or first stated the phenomenon, theory, or law in a form that can later be seen to have been correct. In 1964, when L. Pearce Williams completed his biography of Michael Faraday—a pioneering reconstruction from laboratory notebooks of the research trail of an experimental scientist—this viewpoint still prevailed, as can be seen in Williams's narrative. For Williams, electromagnetic induction was a phenomenon waiting to be discovered by that investigator who would prove clever enough to find the way to it. He noted situations in which "to the modern reader, the principle lying behind the . . . facts is so obvious that it seems impossible that men of the caliber of Babbage and Herschel could have missed it." As for Faraday, himself: "As Faraday tested the consequences of his own ideas and those of Ampère—still holding an open but skeptical mind—his concepts became clearer and clearer until he was led by them to the discovery of magnetic induction." Williams located and quoted the laboratory notebook summary of the particular experiment in which it is clear to a modern reader that Faraday had produced the effect, adding, "The discovery of electromagnetic induction was not only the culmination of a long search; it was the starting point for the brilliant series of experimental researches in electricity which were now to occupy him for almost thirty years."[1]

To write in such a style today would be to brave charges of naïve realism. Some historians, influenced by the rhetoric of the science studies movement, avoid reference to discoveries altogether, preferring the category "construction of scientific knowledge." Even those who retain the word are constrained to respond to the viewpoint explicated in 1981 by Augustine Brannigan, that "we should explain how certain achievements in science are *constituted* as discoveries—and not how they occurred to an individual." According to Brannigan, "Members of society confer the status of scientific discovery on candidate events," by virtue of certain criteria, including judgments about the unprecedented nature of an announcement of such an event, as well about the "substantive possibility" structured by the tradition and the research context within which the asserted novelty has been produced. The status of these elements that constitute the event as a discovery can be determined only by "members of society," and it is, therefore, only they that can "produce and recognize (i.e. explain) discovery." When historians purport to identify the point at which a discovery first occurred to an individual, as though the discovery had resulted from an encounter between the scientist and nature, they are able to recognize the event as a discovery only because that status has already been conferred socially on its subsequent publicly announced version.[2]

That Pearce Williams was able to identify the precise experiment that he identified as Faraday's discovery of electromagnetic induction only because Faraday afterward published this discovery, it was accepted by the contemporary community of scientists concerned with electromagnetic phenomena as the discovery Faraday claimed it to be, and it has continued to be accepted as such, is uncontestable. That does not mean, however, that we should no longer concern ourselves with the question of how and when a discovery occurs to an individual. That the significance of the nascent event can be fully understood retrospectively only from knowledge of its subsequent role as a discovery is merely one of many classes of retrospective judgment that historians must routinely make. Unlike philosophers and sociologists of science, historians need not commit themselves to any transcendent view of the nature of the scientific knowledge that our subjects believed themselves to have discovered and that their contemporaries did or did not recognize as such. We may be equally interested in the processes through which it occurred to an individual that she had made a discovery and those through which her assertions to have done so were judged by "members of society." In later chapters discussing the fine structure of investigative pathways, we shall focus on the nature of the events that constitute discoveries in the minds of the discoverers. Here we shall be concerned with those events that follow the public announcement of such events, the interaction between the discoverer and the scientific community constituting the domain within which it is made, and in particular the effects of their response on the position of the discoverer within that field.

Each of my sample of scientists produced, relatively early in his career, published results that sufficiently impressed the group of more experienced scientists within the field of his investigative activity to mark him as a rising star within that specialty area. In examining the nature of these first auspicious achievements and their effects on the future position of the scientist himself, we might ask whether, as Brannigan asserts, we have nothing more than the judgment of these "members of society" that these products constituted significant discoveries, or whether as historians we may also rank the achievements along some more lasting criterion of value. A further question to be asked is, to what extent did the position within the group that the young investigator already held at the time of this production both make the work itself possible and give the work favorable access to the evaluation it received? In one way or another, in fact, each of these men was already an insider before he reached this turning point in his pathway.

According to Robert Merton, a talented young person who has attained the support of well-placed seniors at influential institutions is apt to become the recipient of "cumulative advantages" over those not similarly placed. When his sponsors assert that the person has the promise to become outstanding, their recommendations have sufficient force to enable the individual to acquire "successively enlarged opportunities to advance his work," including the "comparatively large resources" that leading institutions can offer, and the time necessary to build toward future successes. These advantages can come even when the individual has not yet completed anything extraordinary, if his supporters can attest to his "quality of mind as encountered at short range."[3]

Lavoisier was the recipient of such advantages well before he had lived up to the promise placed in him. Elected at the age of twenty-five to the Académie

Royale des Sciences of Paris as an adjunct chemist, at a time in which his chemical publications amounted to no more than his two relatively minor papers on the analysis of gypsum, Lavoisier probably owed his selection to a combination of his family's connections, the fact that he had impressed older scientists such as Guettard with his talent, energy, and enthusiasm, and the fact that the competition for entry into the Academy at just this time was not especially strong. Membership in the Academy conferred, however, unparalleled advantages in French science. Aside from pure prestige, it offered some facilities for research, but more important, the chance to come into contact daily with the leading scientists of the nation, to share ideas, on occasion to collaborate, and to have access to a powerful and immediate forum in which to present one's work at the weekly meetings of the Academicians. There was little danger that whatever Lavoisier might achieve in his personal investigative ventures would go unnoticed in Paris.

Accustomed to view Lavoisier as the leader of the chemical revolution, and to see him as self-consciously aiming at that role from the beginning of his study of the processes that absorb and release airs in 1773, we may too easily equate the degree of his success and his standing among his fellow scientists with his ability to persuade others to follow him in the overthrow of the phlogiston theory and the establishment of a "new" chemistry. If we adopt that perspective, then, as is well known, Lavoisier appears to have attained recognition only after a prolonged struggle to win over the other chemists in Paris and a still longer time to achieve similar success internationally. The passage in his laboratory notebook in which he is supposed to have foreseen these events in 1773, however, when placed in its immediate context, suggests instead that he believed that the work of earlier scientists, from Hales to Priestley, had prepared the ground for a revolution, and that what he foresaw was only that he would make some further contribution to a movement already under way. In fact, Lavoisier's early publications were not taken to be revolutionary either in content or intent. We should, therefore, evaluate these productions independently of the role he later took on.

One of the privileges of membership in the Academy of which Lavoisier took early advantage was to ask that his work be evaluated by a commission of Academicians. In August 1773, he believed that he had progressed far enough in his study of the elastic fluids absorbed and released that he was ready to publish a treatise on his results, and he asked for a commission to examine it. The task was assigned to the prominent Academician Jean-Baptiste Le Roy, the well-known pharmacist Louis-Claude Cadet de Gassicourt, Trudaine de Montigny, an influential director of the Bureau de Commerce and amateur chemist who had already acted as patron and adviser to Lavoisier, and Pierre Joseph Macquer, the best-known French chemist of the time. The commission did not merely examine the text of Lavoisier's treatise, but asked him to repeat the key experiments in their presence. In their report the commissioners singled out his methods for praise. He had "submitted all of his results to measurement, to calculation and the balance: a rigorous method which, fortunately for the advancement of chemistry, begins to be indispensable to the practice of that science." They were pleased also that, although the phenomena gave rise to "new and bold ideas," Lavoisier had proposed them "with all the reserve that characterizes enlightened and judicious physicists." In contrast to the confidence with which he had identified the air involved in these processes the previous spring as fixed air, he now expressed uncertainty about the nature of this "elastic fluid." The commissioners agreed: "Nothing puts [us] yet in a position to decide whether the combinable part of the elastic fluid of effervescences and reductions is a substance essentially different from the air, or whether it is the air itself to which something has been added or from which something has been removed, and prudence demands the suspension of judgment on that subject." Believing that there were still "many important things to discover about the nature and effects" of this air, the commissioners "exhorted" Lavoisier to "continue this series of experiments so well begun."[4]

Within the Academy, therefore, Lavoisier's first sustained investigative venture attained strong support at the end of its initial phase, not because it seemed to challenge prevailing views, but because it appeared to the commissioners to embody rigorous quantitative methods, such as they already considered essential to the progress of chemistry. Moreover, what he had discovered about the absorption and release of air in combustion, calcination, and reduction both was important and opened up important questions yet to be answered. In short, they saw their younger colleague as a very able practitioner who had quickly joined the forefront in the advancement of the field.

The enthusiastic support that Lavoisier received from his more senior fellow Academicians helped also to spread his reputation quickly abroad. Early in 1774, for example, Macquer wrote to Torbern Bergman, the eminent Swedish chemist, who had just published a paper on fixed air, "We are beginning to work very much here [in Paris] on that curious and important matter. There has just appeared a work on that subject by M. Lavoisier, one of our Academicians, which is very well made and that I believe will satisfy you." Lavoisier himself acted energetically to gain the attention of the international scientific community, sending copies of his *Opuscules chimiques et physiques* to leading chemists throughout Europe.[5] Even here, however, that he could do so as a member of the august Parisian Academy undoubtedly enhanced the odds that the recipients of the volume would pay more attention to it than if it had come to them from someone with more obscure affiliations.

Coming out two years after the publication of Guyton de Morveau's experiments demonstrating that all metals gain weight when calcined, Lavoisier's *Opuscules* placed him at the center of a problem on which many European chemists were beginning to focus their attention. As the Parisian apothecary Pierre Bayen put it in 1774, to explain "the augmentation of the weight of metallic calces" had been for several years "the goal of almost all the chemists in Europe." Lavoisier had advanced the problem by demonstrating that an elastic fluid is also given off when the calces are reduced, and the task ahead, according to Bayen, was now to identify Lavoisier's "elastic fluid." From his own experiments, carried out on the reduction of mercury calx with and without charcoal, Bayen concluded that the air in question was fixed air, a conclusion that Lavoisier had reached a year before, but of which he had since become uncertain.[6]

Thus, those who encountered Lavoisier "at close range" at this time were favorably impressed with the quality of his mind, his energy, and the high level of his experimental work. Because they were themselves influential members of the most prestigious scientific institution in Europe, their opinions placed him in an exceptionally advantageous position, both to continue his work and to assure that it would be received with attention.

There was in the *Opuscules* no call for revolutionary changes in chemistry. Lavoisier was, in fact, far more cautious than Bayen in assessing the effects of his work on the prevailing phlogiston theory. Where Bayen declared that theory "useless,"[7] Lavoisier merely proposed that in reductions with charcoal the phlogiston contained in the latter may enter into the composition of the elastic fluid rather than that of the metal, a view he thought perhaps not "incompatible" with that of Georg Stahl, the originator of the phlogiston theory.[8]

As Ferdinando Abbri has noted, there was in 1775 and for several years thereafter no "Lavoisier cause," no broad debate aroused by his work. His *Opuscules* did make him widely known in Europe, but not as someone claiming an alternative to current trends. The dominant figure at the time was Joseph Priestley, whose discoveries of new airs were then seen as a truly revolutionary event in physics and chemistry.[9] When Priestley came to Paris in 1774, he visited Lavoisier in his laboratory, not as someone he expected to become a rival, but as "my excellent fellow-labourer in these inquiries, and to whom, in a variety of respects, the philosophical part of the world has very great obligations."[10]

"Early achievement in science," according to Merton and Harriet Zuckerman, "may give an enduring advantage by providing both increasingly abundant facilities for research and early access to the social networks of scientists at the research front where information and criticism are exchanged and motivation for getting on with one's work is maintained."[11] Lavoisier's energetic, ambitious nature perhaps would have sustained his motivation even without such special encouragement, but it is also clear that he avidly sought recognition, and this recognition—that his first steps into a question of intense current interest constituted a significant achievement—must have contributed to the intensity with which he continued in the same direction during the next three years. When his ongoing investigative pathway did lead him, by 1777, to contemplate a broad attack on the prevailing phlogiston theory, both the courage required to do so and his ability to obtain a hearing for his subversive views were heavily dependent on the fact that he had by then long been an insider in the scientific community of Paris, respected and admired even by those who resisted his overtures to change their theoretical orientation.

Have we valid criteria for designating historically the outcome of the first phase of the investigative pathway on which Lavoisier had set out at the beginning of 1773 as a significant achievement, independently of the fact that the scientific community to whom he presented it judged it to be so? We can hardly designate it a completed achievement, because he had not solved the basic problem of the nature of the air absorbed and released in the processes he studied. With retrospective insight we can, however, assess more fully than could the commissioners the magnitude of the methodological achievement of those first six months in Lavoisier's quest. They were impressed that he had "submitted his results to measurement, to calculation and the balance," but apparently not fully cognizant of the degree to which the particular way in which he combined these elements surpassed earlier quantitative work in the field. The use of the balance to measure the weight of substances entering into and produced by a chemical operation was not new, and the commissioners seemed to believe that Lavoisier was merely conforming in this regard to the most progressive practices of the time. We can see more clearly that by the time they reviewed his experiments, he had, by combining such measurements with measurements of the volumes of airs produced or released and with a style of extended calculation, already moved beyond the ways in which previous chemists had operated, and that this achievement marked the beginning of a new era in chemistry. During the next two decades Lavoisier's method became so powerful in his hands that it set new standards for chemical experimentation. More than the conceptual framework of his new theory of combustion itself, it was this methodological innovation that enabled Lavoisier eventually to overwhelm the qualitative arguments of those who clung to older theoretical structures, and successfully to lead a "revolution in physics and chemistry." We do not need to have access to some transcendent "truths" about nature to make this judgment, only access to developments that lay still in the future at the time the commissioners of the Academy examined and reported on the work of their young colleague.

For nearly a decade in his early career, Claude Bernard did not enjoy institutional security comparable to that of Lavoisier. Where Lavoisier came from a well-established Parisian family of the minor nobility, was schooled at the leading educational institutions, and used his connections to facilitate his precocious entry into the Academy, Bernard, the son of a poor winegrower and village schoolmaster in the south of France, arrived in Paris in 1833 with no influential patrons. A relatively indifferent medical student, he made no particular impression until 1839, when his surgical talents prompted Magendie to appoint him as his *préparateur.* We have already noted his continued dependence on Magendie for nearly a decade, years in which Bernard acquired no independent position to carry out his research, and no sustained professional income. Despite these deprivations, however, Bernard was no outsider to the Parisian scientific and medical establishment during these years. Magendie had clearly picked him not only as an assistant, but as a favorite student and potential successor. Magendie had had several other students over his long career, but none so promising as the young Claude Bernard. As Magendie's student, Bernard gained access to other support and made further connections within the tight community of Parisian science. For several years he performed experiments requiring chemical analyses in the laboratory of Théophile-Jules Pelouze, next to Dumas the most prominent chemist in Paris. When his lack of a position and income threatened Bernard's future in Paris, Pelouze arranged a marriage for him to the daughter of a Parisian physician, who brought a dowry sufficient to allow him to continue his research.[12]

It seems clear that not only Magendie, but other members of the Parisian scientific and medical establishment viewed Bernard as a young scientist of exceptional promise, that they attributed his early failures to obtain a position to the scarcity of such positions in France at the time, rather than to any inadequacies on his part, and that they acted to protect his future as eventual successor to Magendie. That they did so in spite of the fact that for several years Bernard had few experimental successes, and that the early theories of digestion he published quickly proved inadequate, confirms the insight of Robert Merton that when well-placed senior scientists identify at close range in a younger person, qualities of mind—and in Bernard's case also of the skilled hand—that appear to them outstanding, they are able to promote that person's prospects even in the absence of extraordinary achievements. In this case they were not able to offer the immediate advantage of increased resources, but they were able to offer access, and other forms of recognition. Thus Bernard was able to present papers to the Académie des Sciences as early as 1844, even though he was not elected to membership until much later. A paper he published in 1844 on the spinal accessory nerve, submitted for a prize in experimental physiology given by the Académie, was awarded the prize in 1847 by a jury whose members included Magendie. When he made his first major discovery, of the digestive action of pancreatic juice, in 1848, Magendie immediately arranged for him to be nominated a Chevalier of the Legion of Honor. When the Société de Biologie was founded in 1848, Bernard was elected vice president.[13] Clearly his seniors identified in Bernard the qualities they believed would lead him to future successes well before those successes had arrived. Consequently, when he did produce, in 1848 and 1849, his first two major discoveries, they were quickly recognized as such in Parisian scientific circles.

The contemporary impact on French biologists of Bernard's discovery of the production of sugar in the liver was well stated by Henri Milne Edwards: "For a long time we knew also that the digestion of starchy materials furnishes to the organism a considerable quantity of glucose, and consequently that one could, at first, suppose that all of the substances of that class which appear in the animal economy derive from that source . . . but an unanticipated discovery and one of great importance came, in 1848, to change the ideas of physiologists in that respect, and to show that there is always the production of sugar in the interior of the animal economy."[14] Thus, in accord with Brannigan's criteria, members of the physiological community conferred the status of discovery on Bernard's achievement by recognizing it as novel, even unpredicted, but plausibly connected to the research context within which he had produced it, and of great importance. Historically, however, we can in addition recognize it as a great discovery that has withstood the test of time and has served as the starting point for further research both by Bernard and by successors ever since.

By this time Magendie was nearing retirement, and increasingly Bernard took over his teaching and was able to use the resources of Magendie's small laboratory at the Collège de France for his own research. In 1854 a chair in general physiology was created for him at the Sorbonne, and the next year he succeeded also to Magendie's chair at the Collège de France. From then on Bernard was institutionally, as well as by reputation, at the pinnacle of French physiology. He enjoyed unparalleled forums for promulgating his discoveries and his ideas in well-attended lectures that included frequent experimental demonstrations. Because Paris was still, despite increasing competition from the German universities, the most prominent single scientific center in Europe, the position that he had attained there made him also a major international figure, generally acknowledged as the most successful experimental physiologist of his time.

In the first two years after Hans Krebs left Otto Warburg's laboratory in 1930, he received little help from his former mentor. Fortunately, other contacts he had made enabled him to find clinical positions in which he was also given some time and laboratory facilities in which to begin on his own. During his year at the municipal hospital in Hamburg, his director, Leo Lichtwitz, made special arrangements for him, even though the hospital was not research-oriented. When he moved the next year to the University of Freiburg, a major center for biomedical research, Siegfried Thannhauser not only encouraged him, but gave him the freedom that Warburg had not, to pursue his own research interests. Even in the absence of overt support from Warburg, Krebs benefited from having been his assistant for four years. When he arrived in Freiburg, he was viewed as the person who would introduce Warburg's powerful manometric methods into that setting. He quickly won the respect and friendship of a group of other researchers who included both distinguished senior figures and talented younger men like himself.[15]

In April 1932, the news of Krebs's discovery of the ornithine cycle of urea synthesis spread quickly through the German biomedical community. Thannhauser, an expert on metabolic diseases, immediately appreciated the significance of the work and informed others about it. Within weeks Krebs received congratulations from such leading German biochemists as Otto Meyerhof and Otto Neuberg. Franz Knoop, the editor of *Hoppe-Seyler's Zeitschrift für Physiologische Chemie,* to whom he submitted his first publication on the subject, and whose lectures on biochemistry had been one of the early sources of Krebs's interest in intermediary metabolism, wrote back to thank him "very much for your beautiful paper," assured him it would soon be in press, and added, "I am very anxious to see what else you will bring forth in this direction." Soon Krebs was invited to present his results at symposia in German biochemical laboratories. Finally even Warburg came to realize that his previously underappreciated assistant had achieved something very important, and arranged for him to give a special lecture at the Kaiser Wilhelm Gesellschaft in Dalhem in December 1932. Nor was this favorable response limited to the German biochemical community. In November of that year, Frederick Gowland Hopkins, the preeminent biochemist in Britain, gave a lecture at the Royal Society in which he highlighted Krebs's discovery as an investigation that had "approached on new lines a fundamental problem which for the past sixty years has been the subject of speculation." He praised particularly the precise methods that Krebs had used to solve the problem.[16]

Thus the biochemical community unanimously conferred on Krebs's ornithine cycle the status of a major discovery. No single powerful figure was responsible for this response, although Krebs's position at Freiburg and his previous acquaintance with figures such as Meyerhof and Neuberg when they had all been at Dalhem probably facilitated the quick attention his paper received. That it was a convincing solution to a problem long held to be central to biochemistry, that it conformed to all the expected criteria for such a solution, yet that the form of the solution was also unexpected, all made the discovery both immediately acceptable and exciting. More than seventy years later the ornithine cycle, in essentially the form that Krebs proposed it, remains a central feature of the pathways of intermediary metabolism. The initial judgment of the scientific community that first recognized it as a major discovery has been confirmed by the passage of time, and Krebs's achievement can be viewed as permanent knowledge, not the mere assent of "members of society."

Early recognition had exceptionally dramatic effects on the subsequent career of Hans Krebs. The young clinician who had struck out on his own without strong endorsement from his mentor, with doubts about his own capacity for original research but determined to give it a try, was quickly transformed into a rising star in his field.

Looking back on the situation more than forty years later, he remembered: "I got plenty of recognition, and one thought which was certainly on my mind was that if this is the kind of work you want, please give me the tools and I'll try to do some more, and I think I also had the self-confidence that this had come so easily that I could produce more work of this quality."[17] One year later both the recognition and the confidence became even more crucial, when the Nazi takeover of Germany caused Krebs, as a Jew, to be dismissed from his post at Freiburg. Because he thought so highly of Krebs's work, Hopkins found a place for him in his Biochemical Laboratory in Cambridge. Krebs had other attractive offers as well, but chose Cambridge because of its reputation as a leading international center for biochemical research. He arrived in Cambridge as a young celebrity in the field and thrived in his new environment. In an era in which many young German Jewish scientists were seeking refuge abroad, Krebs stood out for his achievement and was accorded special support. Although his position remained insecure from year to year during his first years in England, he was seen there as of such value that arrangements were made for him to go, in 1936, to the University of Sheffield, where he would have sufficient laboratory space to begin attracting students of his own.[18]

Because he possessed neither long-term institutional security comparable to what Lavoisier had enjoyed from very early, nor the consistent advocacy of a powerful mentor such as Bernard had had in Magendie, early personal achievement was especially critical to the future of Hans Krebs. What he was able to do on his own in two short years with the methodological tools Warburg had given him brought him an international reputation sufficient to survive the political disaster that befell him in his own country, to regain almost immediately the momentum he had acquired at Freiburg, to have access to the leaders in his field, and to work in a stimulating environment where he was soon able to demonstrate that his good fortune in discovering the ornithine cycle was not a singular event, but the harbinger of an ongoing productive investigative pathway.

Seymour Benzer benefited from forms of "cumulative advantage" that enabled him to acquire the support, facilities, and experience necessary to make the transition from physicist to phage biologist without undue pressure to produce quickly results of great originality. He had ample time to make his mark, not because of long attachment to any single mentor or early acquisition of a prestigious position, but because he formed multiple connections with each of the central figures in the strong, informal network of the "phage group." By the time he had spent one year at Oak Ridge, two at Caltech, and one at the Institut Pasteur in Paris, he had associated at "close range" with most of the luminaries in this emerging field, including Salvador Luria, Max Delbrück, Alfred Hershey, and André Lwoff, as well as upcoming younger members such as Renato Dulbecco, François Jacob, Elie Wollman, and Gunther Stent, all of whom formed high opinions of the quality of his mind and his experimental rigor and imagination.

When, therefore, Benzer came upon and began in the spring of 1954 to exploit an experimental system uniquely suited to produce a fine-structure map of a particular region in the genome of the bacteriophage T4, those with whom he was connected in this network responded quickly to the promise of his new project. A frequent participant in informal regional meetings of members of the phage group, he told them in person and by letter of his early progress. He gave his first seminar on the topic at Cold Spring Harbor in July, only a few weeks after obtaining the first preliminary results that persuaded him he would be able eventually to carry the mapping down to dimensions approaching those of the nucleotides of a DNA molecule. Shortly thereafter he wrote a paper summarizing these early results and sent it to Delbrück for comments. Although characteristically critical of some of Benzer's bolder interpretations, Delbrück also expressed his strong admiration for the work. Soon afterward an abstract of the paper appeared in the *Phage Information Service,* an informal mimeographed pamphlet that Delbrück circulated among members of the phage group to keep them abreast of one another's progress. By December the work was already widely known and drawing acclaim from the other members of the group, and Benzer was emerging as a rising star among them.

A comparison of reasons for the rapidity with which Benzer's achievement was recognized with that of Hans Krebs's first major achievement reveals the way in which markedly different styles of community interaction can lead to similar outcomes. Krebs discussed his work on urea synthesis only with local colleagues at Freiburg until he had reached the solution that proved nearly definitive (his first paper did not include citrulline in the ornithine cycle, which was added in a second paper published several weeks later). Communicated through both informal and formal channels, the paper itself quickly persuaded senior leaders in the field of its importance as a completed work. However he may have been regarded by these leaders beforehand, Krebs himself suddenly emerged to them as a highly accomplished experimentalist on the basis of this achievement. Operating within a more open, informal group, which encouraged many personal interactions, Benzer was seen among them as so promising that long before his new work had reached completion it was being recognized as something of great potential importance. By the time he published the papers that fulfilled the promise of these early stages in his investigative pathway, its broad significance was already recognized among that small group of intensely interacting scientists who were helping to forge the new molecular biology.

Meselson and Stahl, too, benefited from their participation in the rapidly communicating network of the phage group. Although not connected in as many ways as Benzer to its other members, they produced their experiment in its "Mecca," the laboratory of Max Delbrück at Caltech. Here, too, word spread so quickly through the network that the significance of their elegant experiment was widely appreciated even before the appearance of the June 1958 issue of the *Proceedings of the National Academy of Sciences* in which they formally presented their results (Figures 4 and 5).[19]

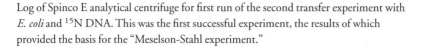

Log of Spinco E analytical centrifuge for first run of the second transfer experiment with *E. coli* and ^{15}N DNA. This was the first successful experiment, the results of which provided the basis for the "Meselson-Stahl experiment."

If we now compare the experiences of our five cases, we can see that each of these scientists was enabled to reach a position of early distinction in part through some form of privileged access. The forms and combinations of access differed. Collectively they included the opportunity to learn how to do science in proximity to an inspired style of leadership, whether through prolonged as-

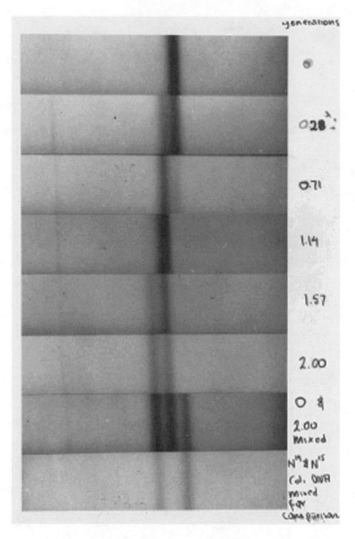

generations

0

0.28

0.71

1.14

1.57

2.00

0 & 2.00 Mixed

N14 & N15 (d. DNA mixed for comparison)

Composite photograph of positives printed from films for centrifuge runs of the second transfer experiment with *E. coli* and ^{15}N DNA. The films were reversed, so that the density increases here from left to right. Top, pure "heavy" DNA taken at time of transfer from heavy to light medium. Fourth film, "heavy-light" DNA taken after one generation. Sixth film, mixture of "heavy-light" and "light" DNA taken at end of second generation. Photo courtesy of Matthew Meselson.

sociation with a single mentor or exposure to several more senior scientists; the opportunity to gain experience with methods of special power and effectiveness; and inclusion at an early stage in institutions or informal group networks affording easy and rapid attention to their potential and actual accomplishments. Our subjects did not benefit equally from all these forms of access, but all benefited from at least some of them. All were in one sense or another insiders before they produced the result for which they first attained major recognition.

None of these individuals, however, reached distinction merely by exploiting these advantages. Each of them responded to what he had received by producing something not only of importance, but of genuine originality, adding his own considerable resourcefulness and imagination to what his seniors had offered him. It was because each added greatly to what he had been taught, because each had something very special to impart along the lines of communication to which he had easy access, that individuals such as these stand out from the many more young scientists who, given similar early privileges, have done more ordinary work or passed along the margins of the history of science. In answer to the question that heads this chapter, therefore, we may state that distinction is both achieved and conferred.

Are the forms of insider access, some of which each of our subjects enjoyed, so essential to a successful investigative career in science that it is next to impossible to achieve distinction without at least some of them? The few well-known examples of outsiders who have done so counter that conclusion, but their rarity suggests that the hurdles which one must cross to do so are very high. Perhaps the most famous case of all is Gregor Mendel, whose great achievement in his experiments on plant hybridization became, decades after his death, the foundation for a new science. That Mendel did not impress his colleagues at the time he published his experiments has long appeared a mystery. That his paper was circulated, and that at least several people who might have been in a position to make it better known did notice it have suggested that Mendel was not the totally isolated figure of the legend. Nevertheless, he was, in comparison to those discussed above, a distant figure, not known at "close range" to those in centers of influence. He was, therefore, not in a strong position to compete for the attention of his peers. Attention being the prerequisite for recognition, it is, therefore, not surprising that Mendel's work was slow to achieve either.

Chapter 6 Maturity and the Dilemmas of the Aging Scientist

It is a widespread belief among scientists, Merton and Zuckerman wrote in 1972, that "the best work in science is done at a comparatively early age." Among the reasons the authors suggest in support of such a generalization, if true, is that recently trained scientists are more likely to be up-to-date on the latest views and techniques available, and to have been trained by a variety of specialists in different aspects of the field, whereas older scientists who have long concentrated on their special interests are likely to have lost touch with developments outside a narrow segment of the field. Although true of all fields of science, they believe that such factors may act most powerfully in the more "codified" sciences, such as physics, in which the empirical content to be mastered by long experience is less dominant, and the capacity to draw inferences from fundamental principles is more extensive. Accordingly, the median age at which scientists "do their most important work," is somewhat lower in physics than in chemistry, and somewhat higher than either of these in physiology and medicine.[1]

In its extreme form this "ideology of youth" presumes that scientists have only a limited store of creative ideas, and that they burn out after this repertoire is exhausted: or, that their creative powers gradually diminish with age and are only partially compensated for by the gradual accumulation of experience. Later in their careers scientists who cannot maintain the quality and originality of their early work are likely to drift more heavily into teaching or administration, slowly diminishing, or ultimately ending their research efforts. "There is reason to believe," however, Merton and Zuckerman assert, that "the general pattern of shifts from research to other roles holds more for journeymen scientists than for the more accomplished scientists. Sociological theory leads us to expect and scattered evidence leads us to believe that the more productive scientists, recognized as such by the reward system of science, tend to persist in their research roles."[2]

Some of the evidence they adduce for this conclusion derives from Zuckerman's study of Nobel Laureates, who, she found, began publishing earlier and continued until later in life—often beyond the age of seventy—than their less distinguished colleagues. Among the reasons to which the authors attribute this persistence is that well-recognized scientists are motivated to continue to live up to the high expectations set for them by their early successes: "What appears from below to be the summit of accomplishment becomes, in the experience of those who have reached it, only another way station. Each contribution is defined only as a prelude to another contribution." Another reason for their continued productivity is that, in accordance with the principle of cumulative advantage, "their earlier achievements in research ordinarily provide them with enlarged facilities for research."[3]

In common with other studies of productivity and creativity in science, this analysis by Merton and Zuckerman tacitly treats the "contributions" and "discoveries" of accomplished scientists as separable, individual acts. If we view them instead as steps along a continuous investigative pathway, then we may find further reasons for the more persistent research trajectories of scientists who have received recognition for major early achievements. These achievements themselves are not isolated successes, but early landmarks along an imaginative, well-chosen investigative venture. Each advance sets favorable conditions for further movement, each problem or sub-problem solved opens up further unsolved problems, each achievement enhances for the investigator the feeling that further achievements lie along the pathway still ahead of her. Moreover, early successes may give her increased confidence, during a later period when she is blocked, that a way will eventually open up to lead her past whatever obstacle she has encountered. There is always a further step to take, and the motivation to continue through a long and productive life may derive at least as strongly from the intrinsic dynamism and unending interest of the route the investigator is continuing to travel, as from the extrinsic rewards the investigator may have accrued from his previous achievements. Although the later advances may seem from the outside less dramatic or fundamental than the early discoveries on which his reputation was based, to the investigator himself they complement, extend, give further depth, breadth, or rigor to what he had earlier found, and may afford a continued personal satisfaction independent of the strength of their impact on other, often younger scientists who have in the meantime taken up problems more attractive to the next generation.

All six of the subjects of my detailed studies were able to sustain the promise of their early successes by maintaining productive research pathways through their later lives (the three living subjects, now entering the years of normal academic retirement, are still actively engaged in experimental projects). The recognition they had received was undoubtedly among the motivations that led them to persist, but their enthusiasm for what they did was clearly derived in large part from the attractiveness to them of the activity itself, and the continued satisfaction they received from their ability to find interesting problems that their previous experiences along their investigative pathways had placed them in positions to take on with steadily accumulating skills and confidence.

Despite the many other claims on his time as a member of the Tax Farm, as a gentleman farmer on his country estate outside Paris, as a very active member of the Academy of Sciences who filled a succession of important administrative posts, and during the early phases of the French Revolution as an engaged reformer, Lavoisier never deviated from the broad outlines of the investigative pathway he had laid down for himself in 1773. Able on the average to allocate only one full day per week to his research, he carried on the remainder during early mornings and evenings. Often interrupted for weeks, sometimes months at a time by his other responsibilities, he invariably returned, as soon as he was able, to the investigative agenda. Among the various types of processes involving the fixation or release of airs that he had originally defined as the broad scope of his investigative goal—including calcinations, combustion, respiration, fermentation, and other unspecified "chemical processes"—he frequently shifted from one to another over the short range, sometimes returning only years later to one taken up and then abandoned, but he never lost sight of the manner in which these facets fit together as one comprehensive set of interrelated problems.

Over the long course of his investigative pathway Lavoisier's motivation for engaging his energies in it evolved as his circumstances changed. At the beginning he was lifted by the hope that he had found at last that "fine course of experiments" that would enable him to make a more conspicuous scientific mark than his previously scattered efforts had yielded. After he had devised relatively powerful methods to pursue this plan, but encountered setbacks for his first explanation of these phenomena, he was probably driven mainly by the desire to find his way out of the theoretical quandary in which he found himself because of his inability to identify the "elastic fluid" involved in the processes he had been studying. Three years later he was still striving to link together his partial explanations for the several processes he had studied into a coherent overall theory, and one can assume that succeeding in a quest that had proven far more difficult than he had imagined in his initial self-assured approach was still his overriding concern.

By the time he achieved this unification of his conceptual structure, in 1777, Lavoisier had formulated a new general theory of combustion that was at once the solution to the problems he had originally taken up, and a challenge to the theory of phlogiston that had previously explained the same phenomena. To win a hearing for his new viewpoint now became a central objective for the next phase of his investigative pathway. That challenge would have appeared to him both more difficult and more urgent than in 1773, when other French chemists had already begun questioning the traditional phlogiston theory, because of the auspicious manner in which Joseph Priestley had in the meantime adapted the idea of phlogiston to explain a whole range of new phenomena made visible by his discoveries of new species of "airs." Curiously, having reached this point, Lavoisier did not for several years press his incipient campaign against the phlogiston theory. It is not clear to what extent he was distracted by extra-scientific matters, and to what extent the delay was due to a grand plan he conceived in 1778 with another Parisian chemist, Jean-Baptiste Bucquet, to repeat a "large number of the experiments carried out up until the present day" in chemistry under the new point of view represented by his theory of combustion. If it had been Lavoisier's intention thus to build a much more comprehensive foundation of chemical knowledge on which later to challenge prevailing views, that goal was ruined by the illness and early death of Bucquet in 1780.[4]

The next stages in Lavoisier's investigative pathway, undertaken in collaboration with the mathematician Pierre Simon Laplace, were, as Henry Guerlac has emphasized, an excursion into chemistry as a "branch of physics," as they sought to measure the amount of heat absorbed or released in various physical or chemical processes, as well as in respiration.[5] Then the unexpected discovery of the synthesis of water gave a new and unanticipated turn to Lavoisier's pathway. After demonstrating this synthesis with Laplace, Lavoisier collaborated with another mathematician, Jean-Baptiste Meusnier, to complete the proof of the composition of water by decomposing it into its two parts. As is well known, these events gave powerful new support to Lavoisier's general theory of combustion, and for the first time leading chemists in Paris began converting to his point of view.

In this phase of his career Lavoisier is ordinarily represented as devoting his main effort to the campaign to prevail over the remaining adherents of the various phlogiston theories still extant. This work was, however, largely literary, expressed in his devastating critique "Reflexions on the Theory of Phlogiston," in the French translation and commentary on Richard Kirwan's defense of phlogiston, and in the reform of the chemical nomenclature in which he and his followers reexpressed in more systematic language the theoretical structure of his theories of combustion and of acidity.[6] In the laboratory Lavoisier was engaged rather in the investigation of further problems made visible by his progress so far—problems that went beyond the outcome of the debates he and his followers were carrying on against the resistant phlogistonists. The discovery of the composition of water led Lavoisier to reinterpret several processes that he had previously attributed to the combustion of carbon alone instead to that of the combined combustion of carbon and inflammable air (hydrogen in the new nomenclature). These processes then led him to devise methods to study the combustion of organic materials such as oils and wax that contained carbon and hydrogen, and these efforts evolved in turn into attempts to determine the quantitative proportions of the two elements in these substances. Afterward he took on the more difficult task, at which he only partially succeeded, of determining the proportions of carbon, hydrogen, and oxygen in sugar, all of which reached a climax in his studies of fermentation. Lavoisier regarded the ultimate test of his new chemistry to be the construction of a balanced equation representing the elementary compositions of sugar and the two products of its fermentation, alcohol and carbonic acid (Figure 6). He did not rest even there, however, but went on in 1790 to further experiments on respiration, a path that he saw extending outward to encompass eventually all the processes of the "animal economy." His pathway came to an end only because the later events of the French Revolution more and more engulfed him, robbing him first of the time and leisure necessary to pursue his interests in the laboratory, and eventually taking his life.[7]

In these late stages of his career we see Lavoisier motivated, perhaps in part

Detail of the painting *Antoine Laurent Lavoisier and His Wife*, 1788, by Jacques Louis David. The Metropolitan Museum of Art, Purchase, Mr. and Mrs. Charles Wrightsman Gift, 1977.

by the desire to continue the exploits that had won him recognition as the leading chemist in Europe, and by his role as leader of what was by now acknowledged on all sides as a revolution, but beyond that, by the same intense curiosity about what lay just beyond in the investigative pathway he had long pursued, which had inspired him to take up these problems in the first place. His pathway had led to a far longer series of "fine experiments" than he could have envisioned when he began, and success in each stage of it had opened up new possibilities. His ever growing mastery of the methodological techniques he had begun to develop in 1773, the ever widening repertory of variations on these methods, the personal financial resources that enabled him to have designed and built for him ever more elaborate, precise instruments and apparatus to carry out his plans, invited him to go on and on, continually exploring the new territory that each of his advances brought into view.

That Lavoisier was able to remain for so long the dominant figure at the forefront of a field of activity that his early successes opened up was due in part to his own energy, foresight, and persistence, but in part also to the absence of strong competitors. The resistance of his opponents to his theoretical claims and to the methods with which he supported them allowed him a clear field for his own continued progress. The fact that he was able to afford equipment far more costly than that of an ordinary chemistry laboratory also brought him a special cumulative advantage—one that Joseph Priestley, who relied mainly on inexpensive, improvised apparatus, decried as unfair and undemocratic.[8]

For a decade following his discovery of the action of pancreatic juice on fats and the formation of sugar in the liver in 1848, Claude Bernard published research papers marking the continuing progress of his experimental investigations at a breathtaking pace. Many of the discoveries he reported were minor, but several were landmarks in the history of physiology. Among those included in the usual list of "major" discoveries credited to him were the effects of sectioning the sympathetic nerve on the circulation and temperature of a rabbit in 1851, regarded as the first step in the elucidation of the function of the vasomotor nerves; the discovery in 1855 that the source of the sugar formed in the liver is an insoluble substance contained in the liver itself, which he named glycogen, and which he isolated in 1857; the analysis in 1855 of the physiological action of the poison curare; and the mechanism of action of carbon monoxide on red blood cells in 1858.[9]

His astonishing productivity during these years can undoubtedly be ascribed in part to the reinforcement of Bernard's earlier efforts by the recognition he was receiving in ever greater measure, by the increased laboratory resources he now commanded as the successor to Magendie, and by his response to expectations that he continue to perform at the high level which had enabled him to attain his position of eminence. At another level, however, we can see that his productivity was linked to the fact that each of these discoveries was not an endpoint, but the starting point for another branch of his ongoing investigative pathway. Each discovery raised as many problems as it solved, and these problems continued to occupy his further attention for as long as his investigative energies endured. Except possibly for the study of carbon monoxide poisoning, for which he believed that he had reached an ultimately satisfying explanation of the mechanism at a molecular level,[10] Bernard never came in his pursuit of any of these problems to a resting point. There was always further to go, and more to do.

The discovery that the liver produces sugar immediately raised two further questions: what is the chemical substance or substances converted to sugar, and where do these substances originate? It was possible, as Bernard quickly realized, that the liver merely condenses sugar brought to it from nutrition, and later releases it, a possibility that he believed he had made unlikely by showing that the liver continues to form sugar in an animal that has fasted to near starvation. For some time he was sympathetic to a proposal from the German physiological chemist Karl Gotthelf Lehmann, that the liver transforms nitrogenous substances arriving in the blood into sugar. In 1855, however, he was able to form sugar in an isolated liver whose arteries and veins had previously been flushed out with water. The source, therefore, was clearly an insoluble substance deposited in the liver itself. That knowledge in turn invited efforts to isolate and characterize the substance. This Bernard succeeded in doing by 1857. In outline, this last step completed the discovery of what has since been known as the "glycogen function of the liver"; but it left much still open for further study. Was glycogen contained only in the liver, or was it more widely distributed? What was the source of sugar in fetal life? These and other questions occupied much of Bernard's time during the next several years.[11] Later in his career Bernard reoriented his viewpoint on the glycogen function of the liver to connect it with his broader concept of the *milieu intérieur*. The problem now became not merely how the organism produces sugar, but how it maintains a constant level of sugar in the blood, a problem that brought to the forefront the nervous regulation not only of the glycogen-sugar reaction in the liver, but of the elimination of sugar from the blood in the kidneys.[12]

Similarly, his unexpected discovery in 1851 that sectioning the sympathetic nerves causes the ear of the rabbit on the same side to become turgid with blood and warmer than the opposite ear raised for Bernard a whole series of new questions. Soon afterward he was able to complement the result by showing that stimulation of the sympathetic nerve reversed these effects. It was difficult, however, to separate the effects on the circulation from those on the temperature, and Bernard struggled for much of the rest of his career to resolve the question of whether all the changes he observed could be attributed to the local acceleration and deceleration of the circulation, or whether the sympathetic nerve also more directly controlled the production of heat in the affected tissues. Moreover, the observation of nervous regulation of local areas of the circulation stimulated him to examine other analogous situations. By 1860 he had been able to show that the salivary glands have a double innervation. One set of nerves increased the circulation through the gland and stimulated secretion, whereas the other set had the opposite effects.[13]

The investigative pathway that Bernard traveled with such intensity during these years thus appeared multiple, in that he kept several seemingly independent lines of research going in parallel. Among the categories outlined by John Ziman (see Introduction), he might be called a "diversifier." The broad range of Bernard's work during this most fruitful period of his career has, however, been interpreted historically only at the level of his published papers. Publications typically separate ongoing investigations into discrete units, and the identification and labeling of Bernard's recognized "discoveries" may partially conceal the interconnections between them. A reconstruction of the finer details of his pathway from his surviving laboratory notebooks might reinforce the view that he was able to pursue several parallel pathways through these years while keeping them mentally separate; but it might also bring them closer together by showing them to be interlocking facets of one or two more broadly pursued goals.

His discovery of the citric acid cycle in 1937 lifted Hans Krebs from the position of an accomplished young investigator in a still emerging field of intermediary metabolism to that of the central figure in a rapidly growing domain (Figure 7). Improvements in the resources at his disposal to continue his work followed soon afterward. In 1938 he was made head of a newly established Department of Biochemistry at Sheffield. The Rockefeller Foundation, which had previously supported him with annual research grants, now began to fund his department at a higher level, in five-year grants, enabling him to expand his operations and to begin to attract a group of students and visitors. By 1940 he had withdrawn from daily work at the bench himself, but supplied the research problems, the close daily supervision, and the ongoing vision that directed the activities of his laboratory. Near the end of World War II, the Medical Research

Hans Krebs in 1939. Photo Courtesy of Philip Cohen.

Council offered to establish a research unit under his direction, and with this enhanced support he was able to assemble a team of workers, some of them passing through, but others of whom stayed with him for many years.[14]

All of these external rewards undoubtedly reinforced the innate discipline and relentless drive that had carried Krebs to the successes that had brought him to his position of eminence. The motivation to push forward, however, came as strongly from the further vistas opened up by the discovery of the citric acid cycle itself. Krebs foresaw immediately that to explore the scope, the significance, and the further details of the cycle was a project that would occupy him for many years. Moreover, each of his other major and secondary discoveries, including the original ornithine cycle, the synthesis of glutamine, and the steps of uric acid synthesis in birds that he had worked out, remained partial solutions. In each case it was clear that further steps in these metabolic pathways remained to be discovered.

Krebs and the students who came to work with him continued to make important contributions to the working out of these and other metabolic pathways. As new methods, such as isotope tracers, and chromatographic and optical methods became available, and many laboratories became involved in a burgeoning field, his role evolved gradually from that of the dominant discoverer in the field to senior statesman, in which he evaluated and synthesized work emanating from numerous other centers as well as his own. Realizing by the mid-1950s that the "main stages of intermediary metabolism are now known," Krebs shifted his attention toward the "mechanisms which govern metabolic rates."[15] Adopting new methods, such as the perfusion of whole rat livers, in which the control of biological processes more closely reflected the physiological conditions of the living organism than in the tissue slices or homogenates then most often used in such studies, Krebs and his team sustained productive investigations of the control of metabolic processes for nearly three more decades, as he moved to Oxford to head the Biochemistry Department there in 1958, and after his official retirement to a smaller metabolic research unit at the Radcliffe Infirmary. Krebs himself maintained the same disciplined schedule, appearing promptly in his laboratory at 8:30 A.M., five and a half days per week, that he had carried on throughout his scientific career. He was still actively involved in the activities of his laboratory until a few weeks before his death, at the age of eighty-one, in October 1981.

According to his own testimony, Krebs was not a long-term planner. From the beginning of his independent investigative career he decided each evening on what experiments to conduct the next day, and he frequently interrupted a line of investigation to try out a new idea that a chance observation or something in the current literature suggested to him. When bogged down on a particular problem, he often turned to another before testing all possible means to make progress against apparent obstacles. He was a resourceful improviser rather than a person who maintained a vision of future goals far beyond his current position. Nevertheless, looked at as a whole, the experimentation he carried out for fifty years has the appearance of one long, unbroken investigative pathway. The problems he examined all fell within his broad initial intention to apply the precise methods he had learned in Warburg's laboratory to elucidate the intermediate steps of metabolic processes. He returned again and again to the same set of problems that he had explored during the 1930s, adding new methods, further details, and deeper insights. The trail never came to an end for him. For Hans Krebs the strategy of moving always from where he had been to a nearby point—adjusting his directions according to changing opportunities but without attempting to leap from the terrain he so thoroughly mastered to more fashionable domains in the life sciences that tended to overshadow the field of intermediary metabolism during the latter half of his career—remained fruitful and productive for a remarkably long scientific lifetime.

Reflecting on the motivations that had led him into and through his scientific work, Krebs wrote in 1970, "No doubt most scientists are emotionally involved in the sense of being deeply committed to their research. I find it difficult, however, to analyze my own emotional motivation objectively and honestly." Believing that, unlike the way James Watson portrayed himself in *The Double Helix,* competitive "racing" had never been important to him, Krebs felt that the powerful new methods Warburg had taught him had protected him from concern that what he set out to do would be overtaken by others. One of his major motives, he asserted, was "simply an insatiable curiosity and the thrill one gets from satisfying this curiosity," which he likened to the playful curiosity of young animals. "But what," he asked himself, "is the nature of the pleasure which the solving of a puzzle provides? Is it the feeling of satisfaction to have been clever enough? Gratification derived from expressions of appreciation, either by one's peers or by those who offer posts, is certainly an inspiring factor." Another driving force in his "earlier days," he acknowledged, was his ambition to justify his choice to become a scientist in the face of the doubts of his father, his mentor, and himself about his ability to "make a success in this field." Finally, "a third force which has not left me to this day, is the justification vis-à-vis those who support me by putting financial resources and facilities at my disposal. This desire to justify a trust is, I suppose, the cause of the same sense of commitment and satisfaction which every worker experiences from the knowledge of a well-done job."[16]

Despite the subjectivity inevitably arising from the fact that here Krebs was looking back on a career that had been eminently successful, his assessment seems to capture much, both of types of motivation common to many scientists, and some particular to his personality and the circumstances of his life. That scientists sometimes mask more egocentric goals under the claim that curiosity has been their primary motivation does not mean that curiosity is to be dismissed as a driving force. Anyone who knew Krebs well would verify that his curiosity about unsolved problems, both large and small, was a genuinely integral aspect of his personality. That he was in the years between his departure from Dalhem and the discovery of the ornithine cycle in doubt of his abilities to do original scientific research, but determinedly ambitious to see if he could, is evident in his behavior during that period. That he responded to appreciation for successes by striving to continue to warrant the trust placed in him was also evident throughout his career. That he felt a deep responsibility to justify the support he received was in part the product of the pre–World War German culture in which he had grown up: one in which reliability and responsibility for one's actions were paramount values.

There were, perhaps, however, other motivations of which Krebs himself may have been less aware. Passing much of his early life in the turbulent outward circumstances of postwar Germany, dismissed and exiled just after his initial successes, uncertain for several more years whether he could stay in England or would end up elsewhere, Krebs found in his disciplined approach to his scientific research, I believe, the stability and order that gave sustained meaning to his life. The second half of his life was as outwardly stable as its first half had been insecure, but by then the structure that his investigative life had given him had become so much assimilated into his person that he had no reason to change what had worked so well for him. In fact, he probably could by then imagine no other way to live. He often maintained in later years that he had been successful not because he was more intelligent than other less accomplished scientists, but because he had refused to be "distracted" from research by the many other possibilities, both professional and social, that come across the path of a scientist. When it appeared in 1965 that he would be forced into retirement, he became seriously depressed, a state from which he was rescued by the arrangements made to continue his work at the Radcliffe Infirmary. For Krebs, to continue his research for as long as he remained competent to do so had become synonymous with continuing to live.

How long can the advantages accumulated by an individual through early successes and persistent productivity last?

Dean Keith Simonton has developed a mathematical model that predicts the longitudinal curve of productivity of creative individuals in a wide variety of activities, ranging from mathematics to poetry. His curves show a steeply rising rate of production beginning with the age at which a career begins, a peak, and a more gradually declining "tail." Along the way he can identify the first major "hit" (normally after a little more than ten years in the field), the "best hit," and the "last career landmark." Although the height and length of the curve vary with the nature of the field and the "creative potential" of the individual, the form remains characteristic. Simonton interprets his curves to mean that individuals have certain creative potential that they consume over a career. In the early stages of ascent they are still absorbing "expertise," in the later stages they are largely drawing on the potential already stored.[17]

Simonton also identifies an "equal odds" rule, according to which the ratio of "hits" to "misses" remains relatively constant for highly productive and less productive individuals, and for the same individuals over the rising and falling phases of their productive careers.[18] As mentioned previously, Simonton is among those who treat each "contribution" as an independent act. For scientists, publications that have wide impact are "hits," those that draw few or no citations are "misses." From the viewpoint of the investigative pathway, however, his curves must have a somewhat different meaning. Although some misses may represent isolated ventures, in the life of the productive investigator they are much more likely to be connecting steps along a broader way. Steps that may not have direct impact on peers may nevertheless be prerequisites to, or at least by-products of the longer course leading to later successes.

Even with such a reinterpretation, the data summarized in Simonton's curves do seem to indicate that, as a general rule, scientists reach a peak of productivity roughly ten years after entering a field, that their most distinguished achievements are likely to take place within twenty years, and that by the time they have been around for forty years they are likely to have passed their last "landmark" event. Somewhere along the way the advantages that accrue even to the highly successful investigator begin to decline relative to those of more recent arrivals.

Where Simonton attributes this decline primarily to the gradual depletion of the individual's creative potential, we might look rather to a variety of circumstances, some related to the intrinsic nature of the later stages of a prolonged investigative pathway, some to changes in the outward circumstances of the lives of scientists, some to the natural consequences of a human life cycle in which aging slows one's pace, even when it does not blunt one's creative edge or diminish the quality of one's reasoning.

Lavoisier did not experience this decline in the usual sense because external events pulled him away from his laboratory at a time in which he was still in his prime. The deterioration we can detect in his last publications on respiration and transpiration clearly resulted from the distractions that prevented him from giving his work the concentrated attention he had formerly devoted to it. No one had displaced him at the forefronts of the fields he had himself opened up. He had maintained so strong a lead over anyone else that it required his followers more than two decades to catch up, to improve broadly on the experimental methods he had introduced, and to supersede significant aspects of the theoretical structure he had built.

Circumstances treated Claude Bernard differently. Around 1860 his health suddenly suffered a sharp decline, from which he never fully recovered (Figure 8). For more than a decade he was able only sporadically to perform experiments in his laboratory. In some of these years he spent many months at his birthplace recuperating, returning to Paris only long enough to deliver his annual lecture courses. In compensation, Bernard spent much time reflecting on his earlier dis-

Claude Bernard as a mature scientist. Photo courtesy of Archives, Yale School of Medicine Historical Library.

coveries and laboratory experience, drawing out from them what he considered to be the fundamental principles of experimental physiology and medicine. He thus maintained his position as the most distinguished physiologist in Paris, but was no longer at the forefront of research in his field.[19]

Even had he not become ill, Bernard would probably not have been able to maintain indefinitely his position as the leading experimental physiologist in an era in which the field was rapidly expanding and growing increasingly competitive. His counterparts in Germany were obtaining large, well-equipped physiological institutes in which they could not only afford more elaborate instruments and apparatus, such as kymographs and other recording devices, precision vacuum pumps, and respiration chambers; but in which they began to train students on a large scale. Karl Ludwig became during Bernard's later years the mentor for many more students than visited Bernard's cramped laboratory at the Collège de France.

Some of the developments in which German physiologists took the lead were based on Bernard's earlier discoveries, but they also went in directions different from his. More oriented toward quantitative measurement than he had been, they sought particularly through their recording devices to find functional relationships between various measurable parameters, such as respiratory movements and blood pressure.

From 1872 until near the end of his life, Bernard's health improved sufficiently to allow him to return actively to his research. Installed now in a somewhat larger, better equipped laboratory at the Muséum d'Histoire Naturelle, he was able also for the first time to gather around him a group of younger investigators, sometimes referred to as his "disciples," who assisted him in the experimental details. About "all this activity," Bernard's biographer, J. M. D. Olmsted, commented that "little was added to Bernard's discoveries that was startlingly new. His work on animal heat, on diabetes and glycogenesis, on asphyxia and anesthesia had been begun years before." The new work did enable Bernard to "add fresh material" in his lectures at the Collège de France.[20]

That Bernard should have added, late in his career, only incremental developments to the topics opened up by his earlier discoveries is just what we might expect of a strongly demarcated investigative pathway that that earlier work had established. It would be hard to imagine how, interrupted for more than a decade, never again robust as he had been in his prime, he could have done better than to return to the problems he knew best, to pursue them where he could beyond the state at which he had left them, even while reinterpreting their meaning in the light of the much thought he had in the meantime given to the broader significance of his previous discoveries. The passing of time did not allow him any longer to startle the field as he had once done, but he was still able to move forward along the trajectory that defined his scientific life. If these were declining years, they were no less meaningful to his personal quest than were those in which he had dazzled contemporaries at the peak of his powers.

The longevity of the scientific career of Hans Krebs was remarkable. Not an early starter, he was thirty years old when he embarked on his independent investigative pathway; yet he was able to pursue that pathway for fifty nearly uninterrupted years. Aware that his mentor, Otto Warburg, who had continued research into his eighties, had isolated himself from younger biochemists, had stubbornly held onto his early theory of cancer long after it had been rejected, and had refused to admit evidence contrary to his theory of photosynthesis, Krebs took precautions in his later years to avoid a similar fate. He remained until near the end of his life an avid reader of the current literature. He traveled to biochemical laboratories throughout the world, where he was regularly the honored guest at informal presentations of the current work in those laboratories. In his own laboratory he depended on younger colleagues and visitors to help him keep in touch with developments in the field.

Moreover, Krebs was able to evade the pattern of Simonton's curves. Although there was a dip in the rate of his publications during the years when his responsibilities as chair of the Biochemistry Department at Oxford claimed much of his time, he regained the pace of his earlier years during the decade and a half in which he again headed a small research group at the Radcliffe Infirmary. How can we account for his ability to maintain almost to the age of eighty a pace of productivity characteristic of a much younger man? One reason was that he was blessed with robust health. Until his brief final illness he rarely missed a day in the laboratory. Nor did the steady discipline with which he had always pursued his scientific life relax in those later years. His own explanation for being able to continue so long was that he never gave up the belief that research was his most important pursuit, and never allowed himself to be diverted from it. Perhaps an additional reason was that his research group in the late years was optimal in size: large enough to be continuously refreshed by younger visitors, small enough for him to be involved with what everyone in it was doing.

The question remains, had Krebs already done his "best work" at an early age? Certainly that for which he is most famous, the ornithine cycle and the "Krebs" cycle, were products of his first seven years in the field. There is no indication that the quality of his later work declined. At the very end we can see a marked distinction, however, between what was for him the culmination of a long investigative pathway, but for most others only a footnote to his earlier achievements. In 1980, the organic chemist Jack E. Baldwin suggested to him at a tea party that the reason that there must be a citric acid cycle may be that the only alternative routes for the degradation of acetate were unable to produce a dehydrogenation. Because the energy released in metabolic reactions can be utilized by cells only through the transport of hydrogen to the electron transport chain, these alternate routes are ruled out.

Inspired by these ideas, Krebs was led to ask the question, "Why have metabolic cycles, such as the citric acid cycle, arisen in the course of evolution?" For many years he had sought explanations for why such a complex cycle of reactions took place. Now he found what was for him a more satisfying answer. Given his understanding of evolution, one should show that this particular set of reactions offered the most efficient, optimal use of the energy released by the combustion of acetate. In collaboration with Baldwin, he wrote a paper outlining his case, which was published in *Nature* in 1981 under the general title "The Evolution of Metabolic Cycles."[21]

In addition to specific arguments for the citric acid cycle, Krebs gave a general argument that whenever a metabolite cannot be directly degraded, but must first be attached to another molecule of low molecular weight, a cyclic pathway is the most efficient pathway, because it regenerates the other molecule, which would be otherwise wasted. The question "of why certain metabolic pathways have evolved can be profitably raised also," he wrote in conclusion, "about noncyclic mechanisms."[22] Excited by these possibilities, Krebs envisioned a "large field to be explored."[23]

In the spring of 1981 Krebs delivered the Dunham lectures at Harvard Medical School (Figure 9). Here he extended his exploration of the new field somewhat beyond that of his succinct *Nature* article, though his explanations of the efficiencies of other metabolic pathways were more tentative than that of the citric acid cycle. His lectures drew a large audience, although attendance did diminish between the first and the last lectures. What impact did the ideas he conveyed to that audience have on them? At least one younger scientist, knowledgeable about the reaction mechanisms that Krebs invoked, suggested to me that Krebs was not up-to-date on the state of that field, and, therefore, that his views were interesting but not compelling. These were, he said "old man's lectures." The large audience came not to hear the latest scientific news from the field of intermediary metabolism, but to honor a historic figure, the architect of

Hans Krebs in 1981. Photo by author.

what had come to appear to them to be the basic foundations of that field, one who had been a leader in it for so long that he had become a legend in his own lifetime.

In two symposia held in his honor in 1980 Krebs had mentioned these ideas as examples of the satisfaction one receives from continuing to do creative scientific work for as long as one remains competent to do so, and as a reason for not considering retirement. For him these ideas did, indeed, provide a climactic vista along a research pathway that he had been following for half a century, one that he anticipated would renew his effort for some years still to come. His death in October 1981 made moot the question whether he would have experienced the disappointment of realizing that others who had not accompanied him on this long pathway would not share his excitement.

Part Three **Episodic Rhythms within an Investigative Life**

Chapter 7 Complicating the
Pathway Metaphor

The metaphor of the investigative pathway suggests a course that, whether straight, winding, or abruptly changing directions, represents a linear progression of research activities. The investigator, having one pair of hands, carries out one operation at a time, and the linear sequence is unambiguous, just as the steps along a route follow one another in an unambiguous order. The reconstruction of the investigative trails of scientists from their preserved laboratory notebook records can readily reinforce this image. The traditional form of bound notebook, in which successive experiments are customarily dated, often provides a linear temporal order corresponding to what the historian can infer is the actual order in which the scientist had pursued the steps of the investigation. The visual image of the traveler conveyed by the metaphor matches, therefore, in a satisfying way the aspects of the literal activity it is used to portray. All metaphors, however, begin to break down when pressed too far, and the pathway metaphor is difficult to extend to imagery encompassing many of the complexities of scientific investigations.

We have already seen that, from the beginning of his career, Claude

Bernard pursued at least two lines of investigation—one on the chemical processes of nutrition, the other on the organization of the nervous system— that he regarded as sufficiently independent for him to keep his records of the experiments on these respective problems in separate series of notebooks. As he successively made major discoveries, each of which invited further work, the number of such separable ventures he maintained increased. Should we envision the same scientist as simultaneously traveling along several research pathways, even though on any given day he probably performed experiments pertinent to only one of them, or should we think of him following a single trail made up of several ongoing "lanes," between which he frequently moved back and forth? Hans Krebs kept only a single series of notebook records during the first decade of his research, implying that he did not see himself as pursuing problems independent of one another. Nevertheless, he shifted frequently from one problem or sub-problem to another, returning sometimes within days or weeks to the earlier one, so that we can see him during his early years as following multiple ongoing problems, sometimes closely connected, sometimes relatively unrelated to one another. The record of his daily *activity* is linear, but his *mental* engagement with each of the sub-problems may have been more nearly continuous, so that what appears from the notebook record a single pathway, alternating frequently between segments of his pursuit of each of the several problems, may represent instead a more nearly simultaneous pursuit in his mind of several ongoing parallel quests.

The idea that most creative people have multiple pursuits Howard Gruber has incorporated in the "organizing concept of a 'network of enterprise.'" According to Gruber, "Enterprises rarely come singly. The creative person often differentiates a number of main lines of activity. This has the advantage that when one enterprise grinds to a halt, productive work does not cease. The person has an agenda, some measure of control over the rhythm and sequence with which different enterprises are activated. This control can be used to deal with needs for variety, with obstacles encountered, and with the need to manage relationships among creator, community, and audience." The enterprises that make up the network of an individual are notable for their "longevity and durability." Once a person has included a project in her network, it is liable to stay there for the remainder of her productive life, even though there may be long periods in which it lies dormant. Because the various enterprises may entail different levels of difficulty or risk, the person may choose at different times to work on a particular one that fits his mood and needs at that point. "Finally, the network of enterprise helps the creative person to define his or her own unique-ness."[1]

Is Gruber's network of enterprise a static, or a dynamic concept? The perma-nent identification of the creative person with his network suggests the former, but Gruber's view of creative work in general as long and continuous growth processes suggests that he may regard the network itself as evolving over time. Gruber does not develop images of the structural forms such networks may take, except to point out that different individuals make different choices between "density and breadth." Some networks may include a broad diversity of projects; in others the particular enterprises may be concentrated within a narrow range of closely related topics.[2]

Can the network metaphor be related to the pathway metaphor? If we take a dynamic view of the network as something the investigator generates and then continually develops over the course of his ongoing activities, then we might think of what Gruber calls the several "main lines of activity" constituting the network as pathways linked together temporally as well as conceptually, so that their intersections represent points in time in which the investigator turns from one to another. Periods in which he pursues more than one simultaneously may be represented by paths that not only intersect, but run together for a while. We may then, perhaps, conjure an image of that person weaving over time the pat-tern characteristic of her particular network.

The problems that arise in elaborating these metaphors would matter rela-tively little to our accounts of scientific investigation if they were merely orna-mental; but I believe that there is a deeper relationship between the representa-tion of scientists as following pathways and our ability to reconstruct their activities in narrative form. Like the pathway, a narrative is necessarily linear, be-cause of the very structure of the prose we must employ to tell a story. When the scientist follows a single line of investigation for a certain length of time, the translation into a narrative is straightforward. When he shifts from one line to another only occasionally, it is not difficult for the story we reconstruct to fol-low him. If, however, the shifts become frequent enough so that we begin to wonder whether our investigator is in fact engaged simultaneously in multiple enterprises, then the question whether to maintain the chronological order of his recorded operations in our narratives, or to separate out the several lines in-stead and to maintain the continuity of each thematic development at the ex-pense of fragmenting the chronology, becomes a serious dilemma.

In my reconstructions of individual investigative pathways I have consciously chosen situations in which the investigator worked alone or with, at most, one or two assistants. I concentrated on the first decade of the work of Hans Krebs not only because he reached during that period of his career his two most prominent discoveries, but because during the later phases of his career, when he directed the activities of a group of students, visitors, and technicians, the situation became so much more complex that it would be impossible to follow at the same level of detail. Then he was able easily to pursue multiple lines of investigation simultaneously, because he could parcel out to different individuals the daily operations necessary to advance each one of them. When the individual scientist becomes head of such a group, then his research trajectory can become too complex not only to visualize clearly through the pathway metaphor, but also to reconstruct in a single narrative thread. How to portray the ongoing activity of such a group at a level deep enough to grasp both the collective and individual pathways followed, the dynamics of the interactions among its members, as well as the processes of discovery, poses a monumental, largely unresolved problem. As in many analogous situations, in probing the nature of investigative pathways through individual cases, we are justified in seeking out relatively simple exemplars rather than those so complex as to defeat our purposes.

David Gooding has devised a diagrammatic representation of the various kinds of moves made by an experimental investigator, in order to map "experimentation as a play of actions and operations in a field of activity," which he calls "the *experimenter's space.*" Using special symbols to represent respectively theories, hypotheses, objects in the conceptual world, objects in the material world, choices, and decisions, with arrows to show moves from one to the next, Gooding represents through the ways in which these are connected the "structure of discovery and construction in this space." Moves that advance the investigation are represented by arrows moving from left to right, those which leave it unchanged move vertically downward. Reiterated operations are represented by arrows slanted upward to the left, which insert "reticulations" into the map.[3]

The "space" filled by such two-dimensional maps is, thus, curiously hybrid. If one simply follows the arrows without regard to direction, one is recapitulating the temporal sequence of operations, a diagrammatic representation of the investigative pathway. The direction of "advance," however, is always to the right, in contrast to my sense of a pathway in which advances may *change* the direction of the investigation.

According to Gooding's interpretation, scientific processes are nonlinear, because they are replete with "bypasses, recycling, repetition, [and] reinterpretation of earlier results." Moreover, his maps are reconstructions of the experimenter's pathway that are only approximations of the original. It is obvious, he asserts, that "the notion of the 'actual' pathway of an experiment is a chimera which language can express but which the reticulations of thought prevent us from accessing."[4] Even the notebook record immediately kept by the investigator is not the direct trace of that path, but the first in a series of reconstructions intervening between the operations actually carried out and the ultimate published version, the latter having gone through several further cycles of reconstruction.

By "nonlinear," Gooding appears to mean that the investigation does not proceed in a uniform direction. From the viewpoint of the pathway metaphor, however, cyclic returns to an earlier point, repetitions, and bypasses are all compatible with a linear process: that is, a process in which the investigator takes one step at a time, so that the successive moves form a single series, or connected chain, whether or not such a chain doubles back on itself.

Gooding has applied his mapping technique primarily to the reconstruction of short but crucial phases in the research of Michael Faraday leading to the discovery of electromagnetic induction. Typically he reproduces in this way operations that occupied Faraday over a period of one or two days. Because of the nature of the experiments, in which Faraday set up experimental systems including electric currents and magnets, varying their relationships to elucidate the nature of the interactions among them, so short a period can encompass a complex series of exploratory moves. It is Gooding's contention that even the immediate translations Faraday made of the movements he had performed and the visual effects he had observed into verbal statements describing them in his notebook required a process of mental reflection that constitutes the first degree of reconstruction of what he had just done. In order to understand such transitions himself, Gooding repeated Faraday's experiments, experiencing at firsthand the distinction between his own motor interventions and visual observations and the written descriptions of them.[5]

Gooding has resourcefully utilized a unique opportunity to penetrate beyond the fine structure of experimental investigation to a kind of ultra structure, in which he can reconstitute activities lasting from minutes to hours. At this intimate level one must agree that the "actual" path eludes primordial description. For Gooding, in fact, the actual pathway is not the goal, for it is the series of reconstructions themselves, the process through which the investigator transforms observed phenomena into "meaning" that most interests him. He uses his experimental maps, in fact, most effectively to diagram the nature of the trans-formations through which these reconstructions are achieved.

How generalizable is Gooding's approach? Is it applicable only to cases in which the nature of the experiments performed and of the record kept reveals a temporal structure on so fine a time scale? Would one find similar patterns spanning longer periods, such as successive experiments performed daily, or even at much longer intervals when the experiments require longer to set up and to execute? Or does the conceptual distance between the "actual" pathway and its first reconstruction represented in the notebook rapidly diminish as the level of resolution of the events we seek to trace becomes less minute? If so, would his experimental maps applied to longer intervals converge on something closer and closer to an "actual" pathway? Would such diagrammatic representations illuminate the nature of such longer pathways as they do Faraday's short pathway segments, or would they become too complex to display meaningful patterns as effectively as an ordinary verbal narrative does?

That the careers of many highly successful scientists fit the pattern of lifelong extended investigative pathways does not imply that all do. There are, in fact, notable examples of scientists who jumped from field to field and achieved major discoveries in each one. Among the most famous of such figures is Louis Pasteur. Gerald Geison has outlined in the following form the research topics that Pasteur took up during his long, distinguished scientific life:

1847–1857   Crystallography: optical activity and molecular asymmetry
1857–1865   Fermentation and spontaneous generation; studies on vinegar and wine
1865–1870   Silkworm diseases; pebrine and *flacherie*
1871–1876   Studies on beer; further debates over fermentation and spontaneous generation
1877–1895   Etiology and prophylaxis of infectious diseases; anthrax, fowl cholera, swine erysipelas, rabies[6]

At first glance it would seem difficult to imagine a research pathway that could lead Pasteur from one to another of such diverse problems, and it is tempting to view him as a kind of universal genius, able to pick up and quickly surpass others in whatever field he chose to work. Pasteur himself claimed, however, that there was an "inflexible logic" connecting his successive research programs, and some of the transitions he made from one to the next seem to bear out his contention.

In 1857, for example, Pasteur explained in his first memoir on fermentation how he had been led from his studies of molecular asymmetry to a "subject of physiological chemistry seemingly so distant from my earlier works." The two subjects were, in fact, he asserted "very directly connected." Amyl alcohol he had found existed in two forms that rotated polarized light in opposite directions. His earlier studies having convinced him that such optically distinguishable forms were also molecularly asymmetric, and that molecular asymmetry could be produced only by "life forces," he inferred that if fermentation produces the two amyl alcohols, then fermentation must, itself, be a process involving living organisms. This view, which he acknowledged to be a "preconceived idea," was "the motive and occasion" to begin the series of experiments on fermentation that occupied him for the next eight years. "But as so often happens in such circumstances," he added, "my work has grown little by little and has deviated from its initial direction—so much so that the results I publish today seem distant from my prior studies."[7]

What might otherwise appear a leap from one field to another becomes, therefore, according to Pasteur's own testimony, the outcome of an ongoing investigative pathway that took an unexpected turn which led by successive stages across an intermediate ground and then went on to raise new problems beyond his original motivation for entering another territory. Similar links led Pasteur from fermentation into the question of spontaneous generation. The connection also between his belief that fermentations require the presence of living organisms which are not spontaneously generated, and his later germ theory of disease was obvious both to Pasteur and to contemporaries. Not all of Pasteur's moves into new problem areas can be explained by such links, however. His studies of beer, wine, and silkworm disease were motivated more by practical considerations than by problems raised during the studies preceding them.

The long experimental career of Ivan Pavlov appears to be divided into two distinct halves. The first, lasting from his first studies under Elie de Cyon until about 1904, and for which he won the Nobel Prize, dealt with the processes of digestion; the second half, an "entirely new line of investigation" of conditioned reflexes, for which he became most famous, continued until the end of his life. Superficially the two topics appear not only unrelated, but in some respects antithetical. In his studies of digestion, Pavlov distinguished "psychic" secretions of the salivary, gastric, and pancreatic glands from secretions caused by the direct action of foodstuffs on the nerves controlling the secretions, and he attributed the former to "decisions" made by the dogs that served as his experimental subjects. The study of conditioned reflexes, on the other hand, reduced what had been regarded as psychic processes to automatic responses. Daniel Todes has shown, however, that the study of conditioned reflexes grew out of a crisis that Pavlov and his co-workers encountered in their studies of digestion, when they tried to apply the distinction they had established between psychic and direct stimulation of the pancreatic and gastric glands to the responses of the salivary glands. Psychic secretion and nervous-chemical secretion turned out in this case to be qualitatively identical. At first, Pavlov attributed the differences in the "psychic" secretions caused by different foodstuffs to the ability of the animal to "recognize" and "judge" the situation, but he soon realized that he was entering psychological territory with which he was not familiar. Inviting Ivan Tolochinov, who had prior experience with current psychological theory and practice, into his laboratory, Pavlov gave him considerable latitude to study the processes of salivary secretion. Tolochinov found by 1902 that the salivary responses to stimulation from a distance could be reduced to law-like regularitites, which Pavlov called at first *conditional reflexes.* During the course of the next several years Pavlov gradually came to the conclusion that he should abandon further studies of digestion and concentrate the resources of his laboratory on the studies of what he came eventually to call *conditioned reflexes.* His complicated decision involved not only assessments of current developments concerning digestive processes that made this field less attractive to him, but also considerations about whether the new topic was amenable to the same large-scale laboratory operation that he had managed so successfully in the study of digestion. Pavlov's shift from one line to another, viewed from close range, appears, therefore, not as a leap into an unfamiliar field, but as a natural turning point along an investigative pathway, leading him in another direction, but incorporating much that he could bring to it from his previous line of march.[8]

Among the subjects of my studies, Seymour Benzer is the most conspicuous example of a scientist who crossed fields. We have seen already that his abrupt shift from physics into phage biology was not so large a leap as it would seem on the surface, for he entered a field that had been partially established by physicists to be more like physics than the traditional biological fields, and he himself began with problems to which he thought he could bring his prior experience in physics.

Benzer continued the project he had begun in 1953 to map the fine structure of the rII region of bacteriophage T4 until 1961, by which time he had succeeded in his goal of "running the map into the ground." His system had become a widely admired model, and many others were now studying it. Suddenly he decided that he had saturated his original aims, and that it was time to move on.

He was now drawn to the question of what "modifications in the cellular mechanism for translating genetic information" could account for the fact that a single mutation in a bacterium could alter the response of a whole subset of phage mutations. To answer this question, he realized, he would have to enter the field of the biochemistry of RNA and the enzymes activating amino acids.[9]

This move would appear, like those of Pasteur, a shift in fields motivated by an unexpected turn in an existing investigative pathway. Before he had got far in this new direction, however, Benzer made a more radical shift. Concluding that the field of phage genetics in which he had so successfully practiced for a decade was becoming too crowded for him to do anything further as distinctive as what he had already achieved, he entered the more classical field of Drosophila genetics, with a new aim: to elucidate the genetic foundations of behavior. After a period of intense study of the existing literature, which probed deeply into the problems of behavior as well as the genetics, Benzer took up the new field and acquired a position of leadership that he has retained for more than thirty years.

Jonathan Weiner, whose book *Time, Love, Memory* follows Benzer's work in this new field, presented his move into it as a fresh start, having little in common with what he had previously done.[10] A closer look at the period of transition would, I believe, uncover bridges crossing from his earlier pathway to the new one. Although Benzer testifies to having known little about classical genetics when he began his rII mapping project, he soon recognized that the analogies between that work and classical gene mapping were very strong, and by 1965 he had absorbed much more knowledge of the field to which he then turned.

In my experience in the reconstruction of investigative pathways I have found repeatedly that the fine structure of the investigation reveals routes across what appear from a greater distance to be large conceptual or experimental leaps. Where we do not have surviving materials to document such routes, it is a reasonable surmise that in many cases the investigator did traverse such connecting pathways, but that they have since vanished from view. We should be cautious, however, about assuming that there must always be such routes. It would be too easy to make of the investigative pathway a concept so malleable that we can stretch it to fit any situation, and thereby deplete its heuristic value.

When we do encounter situations in which a scientist appears to have made genuine leaps from one investigative pathway to another unrelated in any obvious way, and pursued work in the new area with conspicuous success, we might rather ask how that individual managed to circumvent the ordinary disadvantages of entering a field to which her specialized past experience was not applicable. Was there some very powerful generic talent or expertise that she could bring to the new field? Was the new field open enough, or sparsely enough trodden so that he did not encounter others better prepared to solve its problems? Did she quickly invent new techniques or conceptual approaches better suited than the prevailing ones to make progress? Was he so thoroughly blocked in his previous field that it became less productive to remain in it than to endure a new period of apprenticeship to catch up with those already in the new field? Was the person so securely positioned professionally that he was able to obviate pressures to be continuously productive, and free to take up a project whose fruits would be slow to mature?

We should also be alert to the probability that some very successful scientists, even those who remain in one field, have not pursued their research in ways that resemble the pathway metaphor, but have become specialists in opportunism: that is, they are especially adept at picking up a currently fashionable problem, technique, or idea, exploiting it quickly, and moving on to another one. The pathway metaphor can be as useful in identifying those whose work does not fit its patterns as it is in reconstructing the work of those who do.

Chapter 8 Complicating the Episodic Rhythms

Howard Gruber's statement that the creative life is organized into "temporally compact periods within which a given orchestration of effort is played out, and certain projects executed" (see chapter 2), appears on the surface self-evident. The episode on which his study of Darwin concentrated, the period 1837–1839, initiated when Darwin opened his species notebooks and began to think systematically about the species question, and ending when he found in reading Malthus an idea that he later believed had provided him with the powerful concept of natural selection which guided all his further thought about evolution, offers a well-demarcated exemplar of such an episode within a longer creative quest. No matter how singularly and how continuously an investigator may devote a lifetime to the pursuit of a particular pathway, we would expect it to include landmarks along the way denoting the solutions of particular sub-problems within the larger program with which she began, or problems that have arisen during the journey. The careers of each of the individuals I have followed can be divided into at least several major phases, each of which in turn may be further subdivided, as has become evident in some of the previous chapters.

In the case of experimental scientists, the basic "unit episodes" would seem often to be marked by individual publications, or series of publications with similar titles, in the periodic literature. A scientific paper is, in fact, presented as the outcome of an investigation with a discrete starting point and a resolution, even as it represents, as well, a progress report about something ongoing. That is, it appears to mark a compact episode within the larger organization of the investigator's enterprises.

Historical reconstructions of the fine structure of the investigative pathway may, in some cases, identify with greater precision than the published record the beginnings or endings of such episodes. In the case of Darwin, for example, the recovery of formerly missing pages from his species notebooks now allows historians to locate the exact date at which he read the pages of Malthus that elicited the response he later remembered as having given him a theory to work by. More often, however, the record of the fine structure may blur the outlines of episodes that appear compact from a greater distance, revealing overlaps between successive episodes; the coexistence of more than one project at a time; the continuing concern with a project seen from the outside as completed; tentative approaches toward a new project preceding the fuller occupation with it; and gradual transitions between what afterward appear to be clearly distinguishable projects, during which the investigator may not yet realize that he has shifted his direction and focus; interludes during which that which defines a new goal or sub-goal has not yet distinctly emerged from a place it may have held as a subordinate theme in a previous project; and many other complications. Gruber's concept of the network of enterprises would lead us to expect, in fact, that the various projects of an investigator will often not follow one another in a sequence of compact episodes, but will temporally as well as conceptually be intermingled in various ways. His own reconstruction of Darwin's thoughts on the species question revealed that reading Malthus did not so sharply demarcate the end of an episode as Darwin himself later thought.

The diffusing effects of a closer look at the apparent boundaries between episodes can be illustrated at several points during the career of Lavoisier. Between June 1783, when he and several of his associates burned inflammable air in oxygen and obtained pure water, and February 1785, when Lavoisier performed a public large-scale demonstration of the operation to convince remaining doubters, the composition and decomposition of water was his major concern. During this period he also carried out other projects, such as the determination of the combining proportions of oxygen and carbon in fixed air (carbonic acid as he soon after renamed it), but even here the composition of water became an important factor in interpreting the results.[1]

The well-orchestrated demonstrations of the synthesis and decomposition of water appeared, then and now, to complete the construction of the new chemical system Lavoisier based on the general theory of combustion that he had proposed in 1777. By removing an anomaly that had appeared to leave room for the defense of the role of phlogiston, they overcame the resistance of several key chemists who had previously held out or been uncommitted, and although the campaign to persuade the larger international chemical community had just begun, Lavoisier appeared now free to take up in his laboratory new projects that would extend the scope of the principles he had established.

His publications between 1785 and 1789 suggest that these years represent a distinct episode in which Lavoisier devoted himself in his laboratory primarily to applications of his chemistry to processes of nature, more particularly to those in the plant "kingdom." He produced several papers on the composition of three plant substances—spirit of wine (alcohol), olive oil, and wax—and included in his *Traité élémentaire de chemie* his recent efforts to analyze sugar and to establish an equation for alcoholic fermentation.

The detailed reconstruction of Lavoisier's pathway over these years shows, however, that there was no sharp demarcation between the period when his central concerns were his theory of combustion and the composition of water and that when he became primarily occupied with the analysis of plant materials and operations. He had taken up experiments in which he burned spirit of wine, in 1784, in part because, since it was a very inflammable body, to study it was a natural extension of his earlier studies of combustion, but more especially at just that time because he thought that the combustion of spirit of wine formed water. In the aftermath of his discovery of the synthesis of water, he was particularly interested in exploring what processes, both artificial and natural, could produce that ubiquitous substance which had until so recently been thought to be elementary. In his first combustion experiments on spirit of wine the only product he collected and measured was the water.[2]

In May 1785, Lavoisier designed a special apparatus that enabled him to measure both the water and the carbonic acid produced in this process, and he represented the results in terms of his customary balance sheet, comparing the weights of the materials employed to that of the carbonic acid and water formed. Shortly afterward he applied similar methods to olive oil and to wax.[3]

What was Lavoisier's purpose in carrying out these experiments? In the many manipulations of his data required, one of his objectives appeared to be to reconcile his results with the composition of water and of carbonic acid he had recently established. He was also concerned with the question of how much of the water collected was pre-formed in the substances burned, and how much was produced by the combustion. For each case he determined the proportions of the constituents in the original substances: for spirit of wine it was carbon, hydrogen, and water; for olive oil and wax, which he thought to contain only hydrogen and carbon, the proportions of these two constituents (Figure 10). In a draft of the memoir in which Lavoisier wrote up these results, he stated that his goal was "to determine with precision . . . the quantity of inflammable air and carbonaceous material" that such combustible substances contain. The title of his memoir—"The Combustion of the Oxygen Principle with Spirit of Wine, Oil, and Different Combustible Bodies"—however, reflects rather Lavoisier's earlier goal, to extend his general theory of combustion to a broader range of combustible substances.[4] His trajectory over these months suggests that Lavoisier began this phase of his investigative pathway as a continuation of an older one, and came only gradually to realize that the main significance of what he had achieved was to have performed the first quantitative analyses of the composition of plant materials. Once he had come to see this, he went on to make such analyses a central aim of his work. The first three substances he had chosen because they were easily combustible bodies. Once he had fixed on his new goal, he turned to a substance, sugar, very difficult to burn completely, but one of critical importance to a plant process that he had also wished for many years to elucidate, that of alcoholic fermentation. During the next years Lavoisier strove valiantly, with only limited success, to devise methods that would permit him to oxygenate sugar fully to carbonic acid and water.

In this case we can see how an ongoing investigative pathway can be gradually transformed along the way from one project into another. The older project remains embedded within the newly emerging project, and there are no sharp boundaries between them. In the first publications that result, boundaries are partially drawn, boundaries that begin, even to their author, to appear more distinct in later writings. From a greater historical distance these retrospectively drawn boundaries appear to divide the investigator's work into the compact episodes of which Gruber spoke.

In the introduction to his memoir "On the Respiration of Animals," Lavoisier explained in 1789 that the effects and usage of respiration could not be known until "very recently," because it was impossible to know anything about what happens in respiration until a series of prior principles had been understood.

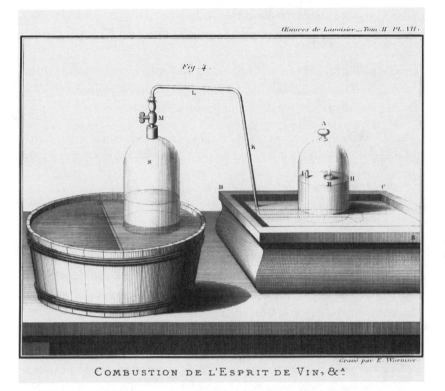

Œuvres de Lavoisier—Tom. II . Pl. VII .

Fig. 4.

Gravé par E. Wormser.

COMBUSTION DE L'ESPRIT DE VIN, &.ᵃ

Apparatus constructed by Lavoisier for combustion of alcohol. The right half was often used for other experiments on combustion and respiration.

These prerequisites included most of his own important previous achievements, from the recognition that caloric is a constituent of gases, to the composition of water. In stating the situation this way, Lavoisier ignored the fact that he had himself presented papers that included discussions of respiration in 1776, 1777, 1783, and 1785: all but the last of them happening before these supposed obligatory preconditions had all been met.[5] Perhaps misled by Lavoisier's retrospective reordering of events, historians have sometimes treated his investigation of respiration as a late episode in his scientific career, as the application of his completed chemical system to the problems of life.

Like some other components of Lavoisier's investigative enterprise, his study of respiration cannot be fitted into any simple episodic structure. He included the topic "the respiration of animals" in his original agenda in the spring of 1773 among the processes that absorb or emit air which he intended to examine. Although he performed experiments on respiration only during three compact periods, in 1776, 1783, and 1790, he was mentally engaged with the problem of respiration throughout the two decades in which he pursued his larger research program.

The earliest trace of Lavoisier's original ideas about respiration are contained in an undated note to himself that he probably composed between the fall of 1773 and the fall of 1774. "Couldn't one surmise," he wrote,

> that the heat of animals is sustained by nothing else than the matter of fire which is disengaged by the fixation of the air in the lungs. It would be necessary to prove that whenever there is absorption of air there is heat.
>
> But isn't the air itself composed of two substances, of which the lungs bring about the separation of one of the two.[6]

With considerable further development, these ideas became the basis for a series of experiments that Lavoisier carried out in 1776 on the effects of respiration on the air in a closed space, into which he introduced birds, the results of which he then published in a memoir;[7] as well as for the calorimetric measurements he carried out in 1783 with Laplace on the heat produced by a respiring guinea pig.[8]

That respiration remained among Lavoisier's important projects even during the long intervals in which he was not actively occupied in its investigation is suggested by the fact that when he recognized that the atmosphere is composed of two principal components, he identified them as its respirable and its unrespirable parts during the time before he was able to give them a chemical definition; that when he did detect some of the chemical properties of the former, he, nevertheless, first named it "vital air" and "eminently respirable air" before he redefined it in 1778 as the "acidifying principle," for which he coined the word "oxygen"; and that when he presented his general theory of combustion in 1777, he included respiration, which he now interpreted as the combustion of carbon in the lungs, as one of his prime examples.[9] In 1784, after he had discovered the composition of water, respiration was one of the most prominent processes that he reinterpreted to include in it the formation of water by the slow combustion of hydrogen.[10]

That in spite of his long-standing commitment to the study of respiration Lavoisier performed so few experiments directly involving it for more than fifteen years may be attributed in part to the scarcity of time that he always had for his experimental work, but probably more fundamentally to his understanding from the beginning that progress in its study was dependent on broader progress in the investigation of the general nature of combustion. By 1789 he believed that that fundamental knowledge was fully adequate to the task, and it was at that point that a project which had until then been interwoven as a subordinate thread within his larger enterprise emerged as his main priority for further ex-perimentation.[11] Only at that level can the study of respiration be said to have constituted the last compact episode within the long pathway of his investiga-tive career.

The only period in the experimental career of Claude Bernard that has been reconstructed in such a way as to enable us to reexamine in finer detail the episodes into which it appears be divided is that from 1843, when he began his first independent investigations, until late 1848, by which time he had just made the two major discoveries that shaped much of the next phases of his work. We can, perhaps, look on these years as one prolonged episode, marking his path-way from apprenticeship to the maturity that arrived with these two achieve-ments. If, on the other hand, we assume that the dozen or so papers he published during these years (excluding here those devoted entirely to the organization of the nervous system) may each represent a "project executed," then we might wish to divide the period into more compact episodes.

Between 1843 and 1846, Bernard published, either by himself or in collabora-tion with the chemist Charles Barreswil, his M.D. thesis and seven other papers, all dealing with aspects of gastric digestion. In 1846 he produced a single paper on a rather different subject, "The Differences in the Digestion and Nutrition of Herbivorous and Carnivorous Animals." In 1847 he published on the role of saliva in digestion, and in 1848 came a rather miscellaneous paper on the chem-ical effects of diverse substances introduced into the organism, followed by the first two papers on his landmark discoveries: the role of the pancreatic juice, and the origin of sugar in the animal economy. From this sequence we might well infer that Bernard concentrated his investigation for three years on the processes of digestion in the stomach, then turned successively to several other problems, all connected to the broad question of digestion as a part of nutrition, but each devoted to some more specific aspect of that larger problem.

The laboratory notebooks record a very different picture. They show that Bernard was preoccupied throughout this period with establishing a general the-ory of digestion. At first it included only the role of gastric juice, but soon it expanded to include the roles of saliva and of pancreatic juice. He proposed sev-eral theories intended to unify these processes, and he labored mightily to pro-vide the evidence necessary to support them. None of these efforts succeeded. The successive theories he and Barreswil proposed were mutually incompatible, and by the end of this period each of them had been rejected by their peers. If one wished to establish within this prolonged period compact episodes, most of them would represent less projects "executed" than projects taken up, abandoned, taken up again, and again abandoned.[12]

The most successful of these papers prior to the two landmark ones that signal the period's conclusion was that of 1846 on the differences between the nutrition of herbivorous and carnivorous animals. This paper was the outcome of a chance observation that Bernard followed up. Noticing that the urine of fasting rabbits was clear and acidic, whereas herbivores normally excreted turbid, alkaline urine, he was able to control the reaction by alternatively feeding rabbits on herbivorous and carnivorous diets. The work attracted considerable attention among his older colleagues in Paris, but it did not mark a discrete episode in his investigative pathway. Rather his experiments on the topic were interspersed among those he continued through this year in his persistent quest to establish his ill-fated theories of digestion.[13]

The discovery of the special action of pancreatic juice, on the other hand, is revealed through the notebooks as occupying an extremely compact period. Although he had tried, apparently several times over the preceding years, to collect pancreatic juice through the pancreatic ducts, the animals had not withstood the operative procedures. Finally in the spring of 1848 his surgical skill enabled Bernard to insert a canula in the duct with so little exposure of the delicate pancreas that he could obtain the pure juice in abundance for several days. After that technical triumph he was able to establish within little more than a month that the pancreatic juice has a special, hitherto unrecognized digestive action on fats.[14]

In May of that same year Bernard undertook a search for the place where sugar ingested disappears within the animal "economy." He relied on two basic methods—subjecting sugar to the action of ground tissues, such as lung or liver, to see if they would cause the sugar to disappear; and withdrawing blood from various arteries and veins to find out if the sugar disappeared while passing through a particular organ, or whether, on the other hand, it might disappear in the tissues in general. Until the end of June he obtained only mutually contradictory results. In retrospect we can see that he could not have succeeded in this project, because his general idea that the sugar disappears at some point in the circulation was wrong. On July 3, however, he recorded the unexpected observation that a dog that had subsisted for eight days on a diet that excluded sugar or starch contained sugar in the blood of its portal vein. Following up this discovery with a burst of intense experimentation, Bernard was able by the middle of October to demonstrate convincingly that the liver secretes sugar into the bloodstream, regardless of whether the animal is fed sugar.[15]

Both the discovery of the action of pancreatic juice on fats and of the production of sugar in the liver took place within very compact episodes along Bernard's long investigative pathway. They were not, however, the execution of the projects that Bernard had planned. It seems evident that his main intention in devising the surgical technique that enabled him to collect fresh pancreatic juice had been to continue his earlier study of the effects of pancreatic juice on meat and on starch. The single test of its action on fats that he interspersed among those on meat and starch suddenly changed the course of his experiments. Similarly, his project to find out where sugar disappeared was transformed midway in its execution to a project he had not anticipated: to establish the source of the sugar not traceable to the diet of an animal.

In Bernard's early pathway we find, therefore, that what appear in his publications as separable projects each terminated by the publication of a paper were not aligned in such an orderly sequence in his daily laboratory work: that he returned again and again to pursue questions supposedly settled in these papers, mainly because further observations conflicted with what he thought he had established, and that ultimately most of these efforts ended rather in failure than in execution. His first, moderate success, the distinction between herbivorous and carnivorous urine, was rather a by-product of his investigative activities than the completion of an intended project. The two very auspicious successes of 1848 that transformed his career were both outcomes of his imaginative response to results that redefined along the way the project he had recently undertaken, rather than the completion of what he had begun.

The list of publications of Hans Krebs during the five years between his discovery of the ornithine cycle in 1932 and that of the citric acid cycle in 1937 similarly presents the appearance of an orderly sequence of episodes, each devoted to the solution of a particular problem, each discretely defined, but all falling within the broader category of establishing steps in the pathways of intermediary metabolism. In 1932 he published, with his assistant Kurt Henseleit, three papers on the urea cycle, the first two being preliminary reports, the third being a much longer paper giving fuller experimental details underlying the conclusions reached in the first two. The next year he published two papers on the "metabolism of amino acids," dealing with the deamination reaction generally regarded as the first step in their decomposition. After his move to Cambridge later that year, he published no reports on his current original research until 1935, when he added another paper on the decomposition of amino acids that represented a further pursuit of the problems discussed in the preceding papers. Shortly afterward he published a single paper, "The Synthesis of Glutamine from Glutamic Acid and Ammonia, and the Enzymic Hydrolysis of Glutamine in Animal Tissues." Each of these papers appears as a bounded investigation, each with a well-defined starting point, each set leading to a significant conclusion, all falling within the area of "nitrogen metabolism," one of the generally recognized sub-fields of intermediary metabolism. At the beginning of 1936 Krebs published, in collaboration with two of the students he had advised at Cambridge, two papers in another branch of nitrogen metabolism, the synthesis of uric acids in birds.[16]

Later that year his papers indicate a shift of interest from nitrogen metabolism to the oxidative phase of carbohydrate metabolism. The first of these, published in *Nature* in August 1936, was a bold scheme proposing a set of three dismutation reactions that Krebs claimed to be the primary steps in the oxidation of pyruvic acid in living cells. Early in 1937 he published, together with his Ph.D. student William Arthur Johnson, two papers treating, at a lower level of generalization, some reactions of the ketonic acids that he thought might play a part in intermediary metabolism. In the fall of that year came the landmark "The Role of Citric Acid in Intermediate Metabolism in Animal Tissues," which provided the definitive scheme for the oxidative phase of the decomposition of carbohydrates.[17]

The reconstruction of Krebs's investigative pathway over these years presents a more mixed picture, verifying that some of these papers do represent the outcomes of compact episodes of intense concentration, with well-demarcated beginnings, and ending with the "execution" of the project begun. Others, however, represent projects pursued intermittently, projects that began as parts of, or unexpected turns in other projects, projects whose definitions changed as they proceeded, and conclusions that represented only partial solutions to the problems undertaken. Almost all of the projects Krebs pursued at Cambridge from 1933 to 1935, and at Sheffield after his move to that university in the fall of 1935, represent the continuation of or return to problems he had first taken up during the two years at Freiburg that preceded his forced departure from Germany in the early summer of 1933.

The discovery of the ornithine cycle is perhaps the most striking example of a project with a very distinct beginning, pursued consistently over a compact period that ended with a dramatic solution to the problem originally posed. Freed finally from lingering obligations to investigate problems of interest to

Warburg in the early summer of 1932, and realizing that his working conditions at Freiburg favored the choice of an ambitious research project, Krebs deliberately chose the synthesis of urea as a strong challenge to his general plan for applying the methods he had learned from Warburg to the problems of intermediary metabolism. Without a clearly defined hypothesis of his own, he set out to test the various intermediate steps that earlier investigators had proposed. After performing a series of preliminary experiments for two months, he turned the daily experimental work over to Henseleit, whom he supervised closely. Two months later, the unexpected ornithine effect gave new direction to the investigation, and four months after that, he reached the cyclic solution to the question with which he had begun.[18]

The study of the deamination of amino acids, on the other hand, did not begin as a distinct project. Rather, because the standard view of urea synthesis had been that the source of the ammonia in urea is the deamination of amino acids, Krebs turned to this subject as an integral part of his study of the former question. Only after he found out unexpectedly that amino acids in general were deaminated especially in kidney tissue, whereas urea was formed in the liver, was his study of deamination transformed into a separate branch of his investigative pathway.[19]

Early in 1933 Krebs consciously added a new project to his network of enterprises, when he set out to study the very broad question of how foodstuffs are broken down in metabolism. This project turned out, however, to be itself a network of smaller projects. Having no developed views of his own on the subject, he began by testing various hypotheses that were generally discussed in the field, or that he found in the current literature. Most of these were related to carbohydrate and fatty acid metabolism, to the connections among a group of dicarboxylic acids—succinic, fumaric, malic, and oxaloacetic acid—that previous workers had shown to have strong effects on tissue respiration, but whose connections to the major foodstuffs remained uncertain, and to certain other substances, such as pyruvic and acetic acid and the "ketone bodies," long suspected to play central roles at the intersections of the decomposition pathways. Because his methods allowed him to perform numerous experiments rapidly, sometimes simultaneously, with the sets of ten Warburg manometers he deployed, he was free to try many such ideas without spending large amounts of time on either their preparation or their execution. Some of the ideas he disposed of after a single negative result, others he pursued for weeks or months. Problems and subproblems were nestled intricately within one another.[20]

When he moved to Cambridge, Krebs continued in the general direction he had begun at Freiburg. He turned up some interesting leads and tested several original ideas of his own for potential pathways. None of these efforts led him very far. Concluding by mid-November that he was "up against a wall," he gave up the whole endeavor for more than two years, returning instead to problems of nitrogen metabolism, where he had earlier been more successful.[21]

Early in 1934, Krebs took up, with Hans Weil, another German expatriate who had come to work with him in Cambridge, a project to study the formation and decomposition of uric acid in birds. This process was equivalent to the formation of urea in mammals, and represented for Krebs the return to a project he had originally given to a student in Freiburg, Theodor Benzinger. As usual, they began by testing in Krebs's experimental system reaction schemes proposed earlier by others. Within a few weeks Krebs believed he had identified an alternative pathway for the decomposition of uric acid, and prepared a paper on the topic. The paper was never published, however, and when Weil left Cambridge later in the year, Krebs left the problem unsolved. During the late spring and summer he took up various problems suggested to him by papers appearing in the most recent biochemical journals, including especially one by Albert Szent-Györgyi on the role of succinic acid in tissue respiration. That direction might have brought him back to his general concern with the breakdown of foodstuffs, but he seemed unwilling at the time to pursue it further. Another lead from Szent-Györgyi's group, on the effects of ammonia on the formation of ketone bodies, he also took up for a while before turning the project over to another student at Cambridge.

On his return from his August holiday in 1934, Krebs went back rather abruptly to two problems on which he had been occupied just before he was excluded from his laboratory in Freiburg fifteen months earlier. One was to characterize the enzyme that was responsible for the deamination of amino acids he had observed then in kidney tissue. Originally he had used racemic mixtures of the amino acids he tested. When he began to test the two optical isomers separately, he found that the non-natural isomers were deaminated more actively than the natural ones, and came to suspect that there were two classes of amino acid oxidases responsible for the actions on the two classes of isomers. The other project was to explore the anomaly he had noticed when he included two dicarboxylic acids, glutamic and aspartic acid, in the deamination study. Whereas the amount of ammonia present increased in the other cases, it decreased when either of these two acids was present, and when he added ammonia to the medium it was absorbed. This effect led him to suspect that each of these acids

formed nitrogen compounds, in the first case glutamine, in the second asparagine.

Through most of the rest of 1934 and the spring of 1935, Krebs concentrated
on both projects, switching back and forth frequently, but with the clear sense
that he would stick with them to resolution. In the case of the two dicarboxylic
acids, glutamic acid and glutamine gave more decisive results than the aspartic
acid asparagine pair, causing him to drop the latter and focus entirely on what
became the study of the synthesis of glutamine. In the study of deamination he
was able to extract and describe the properties of the enzyme that acted on the
non-natural isomers, but did not have similar success with the enzyme presumably required for the natural isomer.

Krebs completed and wrote up papers on these two projects within a few
weeks of each other in the early summer of 1935. The deamination paper, although a success, was a relative disappointment to him, because he had hoped
to characterize the enzyme for the natural isomer, which would have fulfilled an
important cellular function, and he could think of no compelling reason for the
existence of an enzyme that acted on substances not normally occurring in the
organism. The discovery of the synthesis of glutamine, on the other hand,
opened up a new vista in intermediary metabolism. Although his efforts to extend the pathway beyond this single reaction step he had established went
nowhere, he had the satisfaction of establishing a hitherto unknown sequence,
still a rare event in the emerging field of intermediary metabolism.

If we look at these two projects as episodes in Krebs's career, we see that they
overlapped nearly completely in time, that the study of glutamine synthesis did
not begin as a distinct investigation, but quickly emerged from that of the deamination inquiry as an anomalous result, at first involving a class of two amino
acids, which Krebs later narrowed to one of them. After the two projects had acquired distinct boundaries, he pursued each of them intensely, and concluded
both of them within one compact period of time.

In January 1936, Krebs performed a series of experiments on the formation
of amino nitrogen from acetylalanine and from acetylglutamic acid, compared
with the formation from pyruvic acid and ammonia. These experiments were
intended to test a hypothesis originally put forward by Franz Knoop, that the
acetyl amino compounds were intermediary steps in the formation of amino
acids from keto acids. Although Krebs had earlier been concerned primarily with
the deamination of amino acids, he had made a few isolated attempts to examine the reverse process, so that his new venture can be seen as a resumption of

his previous investigations within the domain of nitrogen metabolism. Turning next to the first stage of Knoop's reaction scheme, the formation of the acetyl intermediates from keto acids, Krebs tested several of the latter together with ammonia. Along the way, however, he noticed that the ammonia did not appear necessary to the reaction that was taking place, which produced CO_2. With that observation he left behind the question of the formation of amino acids, shifting his focus to various reactions of the keto acids that might produce anaerobic oxido-reductions by molecular rearrangements, a class of reactions known as dismutations. His problem shifted again several times in the spring, appearing for a time to fix on pyruvic acid, long known to play a pivotal role at the intersection of the anaerobic and aerobic phases of carbohydrate metabolism, but then broadening out again to include reactions between pairs of keto acids, particularly prominent among them being acetoacetic and ketoglutaric acid. Influenced again at this point by the latest work of Szent-Györgyi on the role of the dicarboxylic acids in cellular respiration, Krebs began to examine whether succinic acid was formed in these dismutation reactions, and found that it was, indeed, among the products of some of them.

Encouraged by the evidence that he had produced for the occurrence of these reactions, Krebs submitted to *Nature* in July 1936 a preliminary paper, "Intermediary Metabolism of Carbohydrates," in which he made the bold claim to have identified the "primary steps" in the oxidation of pyruvic acid in living cells, steps which, he added, "proceed also in the absence of oxygen." Although understated, the scheme that he presented would, if confirmed, have linked the recently elucidated pathway of glycolysis and the formation of the ketone bodies into a comprehensive mechanism for the anaerobic and aerobic phases of carbohydrate metabolism.

In this episode, therefore, we see a problem that Krebs took up as a new facet of what had been his principal domain of research in nitrogen metabolism transformed, by a chance observation, into a problem that resuscitated the quest he had previously abandoned to follow the decomposition processes of the nonnitrogenous foodstuffs. Then, within the compact period of several months he was able to "execute" the newly defined project sufficiently to publish the outlines of a potentially major contribution to that field.

This achievement, however, proved ephemeral. As he continued experiments along the same line, Krebs was able to produce evidence for a growing family of analogous dismutation reactions, including several whose products included citric acid, which might play a part in the oxidative decomposition of carbohydrates; but he could not prove conclusively that any specific sequence of these reactions constituted the main pathway. During the fall he delivered several lectures on the progress of his work, in each of which he had to acknowledge that his views on the subject remained provisional and incomplete. This situation continued into the spring of 1937.

During the same period that Krebs focused his attention largely on these problems, he pursued intermittently, in collaboration with Francis J. W. Roughton at Cambridge, a project to identify a further intermediate in the ornithine cycle of urea synthesis. That there was at least one more missing intermediate had long been obvious to Krebs from the fact that the conversion of ornithine to citrulline required three molecules to come together simultaneously. He and Roughton believed that the compound carbamino-ornithine might fill the gap. Because that compound was unstable under physiological conditions, Krebs could not test it in the same way he tested other metabolites, by adding it to the medium of a tissue slice. One had to provide indirect evidence in terms of the equilibrium constant of the compound, using methods with which Roughton was deeply familiar. By the beginning of 1936, the collaborators thought that they had acquired sufficient evidence to write up their results. After consulting with an expert colleague, however, Roughton accepted the opinion that the compound whose physical properties matched their needs was, in fact, the wrong isomer of carbamino-ornithine, and the solution that they had thought they had reached collapsed. Accepting this verdict, Krebs gave up further attempts to find the intermediate that he supposed would complete the ornithine cycle.

Sometime during the third week of April 1937, Krebs came across a paper by Carl Martius and Franz Knoop outlining a pathway for the decomposition of citric acid. The reactions citric acid → cis-aconitic acid → iso-citric acid → oxalo-succinic acid → α-ketoglutaric acid → succinic acid then connected with the dicarboxylic acid series long expected to play a central role. Having tried unsuccessfully at various times during the past four years to connect the decomposition of citric to other metabolic reactions, Krebs understood the significance of this sequence immediately, and set out to test whether it could be demonstrated to take place in living tissue. Soon he devised a cyclic mechanism that extended these reactions to the resynthesis of citric acid. Within six weeks he had established an interlocking set of experimental results providing convincing evidence that the citric acid cycle existed, and that it operated at a rate sufficient to account for the main pathway of the oxidative decomposition of carbohydrates (Figures 11a, 11b, 12).

If we regard the experiments he performed after seeing the paper by Martius

Laboratory notebook pages by Hans Krebs showing first experiment on citric acid carried out after he had read the paper by Martius and Knoop.

and Knoop as a fresh start, then these six weeks constituted a remarkably compact episode in which Krebs began and completed the project that proved to be his most important achievement. Because, however, he was able to reorganize and include in this work several elements from his earlier efforts to identify a series of dismutation reactions that would, if correct, have served the same function that the citric acid cycle does, we can also regard these six weeks as a climactic culmination of a much longer episode that had begun in Freiburg in the

90

2) *Aerobic experiment*

Medium: 3.5g *Muscle* + 15 ml NaHCO₃ 1.3% + 25 *serum* + 50.0 *phosphate* $\frac{4}{10}$ = 120 cc

End experiment

	70	71	72	73	74	75	76	
brain		0.3 $\frac{n}{10}$ NH₃	}→	0.1 $\frac{n}{1}$ *reduct* →	~	~		
oxide	—	··	0.1 $\frac{n}{2}$ *white*		→		~	
10' *equilibri·*								
KO₂								
5'	-250	-77	-147	-945	-88	-99	-101	
5'	·	-50	-130	-4945	-82	-53	-92.5	10'
5'	-186	-205	-89	-	~	~	~	
5'		-135	64/-158	-25	-41	-27	52	20
5'	-187	-6	-49	-18	285	-20	-38	·
5'	·	-5	30	-18	-36.5	-25	-51	35'
5'	-161	-4	-18	KHₐ	*for serum*	*anhaltend*	*for serum*	·
5'	·	-35	-18					
5'	-93	-4/-1185	-13/-534					
		ext· defstly						
KO₂	10 -53	53/281	221/228	238	111	404	366	
factor	2 -	2.35	32	3.30	290	2.93	2.35	
only		135	74	78.6	322	117	112	

spring of 1933, broken off in Cambridge late in the same year, and been resumed in the spring of 1936 when a project to study a potential pathway of amino acid amination changed course and led Krebs back to these previously interrupted projects.

The patterns that we have seen in analyzing the episodic rhythms in the research careers of Lavoisier, Bernard, and Krebs are what we might expect according to the view that experimental scientists typically follow investigative pathways that lead them step by step from where they are at any given time to-

-2-

In this cycle "triose" reacts with oxaloacetic acid to form citric acid and in the further course of the cycle oxaloacetic acid is regenerated. The net effect of the cycle is the complete oxidation of "triose".

The conversion of citric into oxaloacetic acid passes through the following intermediate stages:

$COOH \cdot CH_2 \cdot C(OH) \cdot CH_2 \cdot COOH$ $\quad\quad\quad COOH$	citric acid
$COOH \cdot CH_2 \cdot CH \cdot CH(OH) \cdot COOH$ $\quad\quad\quad COOH$	iso-citric acid (Wagner Jauregg and Rauen²)
$COOH \cdot CH_2 \cdot CH \cdot CO \cdot COOH$ $\quad\quad\quad COOH$	oxalosuccinic acid (Martius and Knoop²)
$COOH \cdot CH_2 \cdot CH_2 \cdot CO \cdot COOH$	α-ketoglutaric acid (Martius and Knoop²)
$COOH \cdot CH_2 \cdot CH_2 \cdot COOH$	succinic acid
$COOH \cdot CH : CH \cdot COOH$	fumaric acid
$COOH \cdot CH_2 \cdot CH(OH) \cdot COOH$	l-malic acid (Green³)
$COOH \cdot CH_2 \cdot CO \cdot COOH$	oxaloacetic acid

The intermediate stages in the synthesis of citric acid are still obscure. A probable intermediate and immediate precursor of citric acid is cis-aconitic acid. This acid yields rapidly citric acid in muscle, liver or testis. About 12 mg. citric acid can be formed from cis-aconitic acid by muscle per gramme

Earliest schematic representation of citric acid cycle, drawn by hand by Hans Krebs on the manuscript submitted to *Nature* in June 1937. This paper was not published. The first publication in *Enzymologia* contains a similar representation. Subsequent representations of the citric acid cycle, known also as the Tricarboxylic Acid cycle and the "Krebs cycle," integrated all of the known steps into a single circular diagram.

ward points that they can only dimly define in advance, that confront them with obstacles which cause them to shift directions, and that open up new opportunities along the way. If sometimes the investigator is able to begin a project and follow it to completion without serious diversion or interruption, that is as expected as it is that at other times the project will metamorphose into something different from what it was at the beginning, that the investigator will often be pursuing more than one project at a time, or that she will sometimes abandon a project, to which she may never return, or which she may resume when some new condition rises to open the way previously blocked or obscured. Some of these characteristics of the research trail are explained by the long time required for the investigator to acquire expertise and the advantages she attains by exploiting prior experience rather than starting over in a new field. They are further explained by the structure of scientific problems themselves. The investigator does not normally encounter a series of linearly related problems lending themselves to an orderly progression in which he can systematically take up one after the other. Rather, any problem that an investigator pursues is part of a larger problem, and has, in turn, nestled within it sub-problems, each of which may include a structure of further sub-problems. There is often no single logical order in which these intricately linked problems ought to be taken up, so that contingent circumstances and encounters instead have much room to influence the choices the investigator may make. The factor that probably most strongly perturbs the orderly succession of problems, however, is the inevitability, in original scientific research, that the investigator will come upon phenomena that are unpredictable until they appear. It is to that subject that we will turn in the next chapter.

Chapter 9 Predictability and Unpredictability

The French molecular biologist François Jacob has forcefully evoked the unpredictable nature of scientific research: "Research is a process in which one can never say how it will evolve. The unpredictable is in the very nature of the scientific enterprise. If what one sets out to find is truly new, it is by definition unknown in advance. There is no way to tell where a given domain of research will lead. . . . One could even say that in fundamental research, if there is not a good dose of uncertainty at the beginning, there is no chance that it deals with an important question."[1] Jacob attributes this unpredictability not to the nature of science alone, but to the general human condition. All our actions are directed to the future, he writes, but we can never know what will happen even in the next moment. Jacob views the unpredictability of science not as a drawback, but as reason to admonish policy makers to avoid constraining scientists to solve problems laid out in advance: to allow them instead the freedom to follow the unpredictable directions in which their research may lead them.

Hans-Jörg Rheinberger has sought to narrow the focus of the unpredictability of science of which Jacob has written, by locating its primary source within the nature of the experimental systems employed by scientists in their research. A researcher, according to Rheinberger, does not deal with isolated experiments intended to test a particular theory, but begins by setting up a "whole experimental arrangement designed to produce knowledge that is not yet at his disposal." Modern experimental biologists, especially, he points out, are wont to begin discussions of their work by describing their experimental systems. The appropriateness of the experimental system devised by the scientist to the problem taken up is the first prerequisite for making progress. Once established, however, the experimental system tends to "take on a life of its own." It becomes, as Mahlon Hoagland has commented, a machine for "generating surprises." The system must have sufficient stability to reproduce results, but be open enough also to generate unexpected events, events that, as Rheinberger writes, "may induce major shifts in perspective within or even beyond their confines." In his view, these characteristics of the experimental system often override the original intentions of the scientists who have designed them. The elaboration of the capabilities of an experimental system "requires familiarity with the system. This process takes time, which helps to explain why experimenters, once they have established their experimental network[,] often stick to it in almost symbiotic fashion. But once the system has become familiar to those who 'inhabit it,' its own momentum may take over. It may force the researcher and research into a kind of internal exclusion. The more the experimenter learns to manipulate the system, the better the system comes to realize its intrinsic capacities: it starts to manipulate the researcher and to lead him or her in unforeseen directions."[2]

Rheinberger has illustrated this point of view with a convincing case history taken from the research of Paul Zamecnik and his colleagues at the Massachusetts General Hospital in Boston. In the 1950s Zamecnik set out to distinguish protein synthesis in cancer cells from that in normal cells. Unable to detect significant differences, the group soon turned to the more general study of protein synthesis in surviving normal tissue slices. To achieve greater discrimination they decided to "take the cells apart," a goal that they achieved by means of an experimental system that enabled them to separate particles of different size contained in a homogenate of liver tissue by fractional centrifugation in an ultracentrifuge. In the early stages of their project they defined what they hoped to find in the traditional terms of biochemistry: to isolate enzymes and identify metabolic reactions. What they were eventually led by the evolving characteristics of their experimental system to discover was a fraction, called first "soluble-RNA" and later "transfer-RNA," whose function was defined not according to the tenets of biochemistry, but to those of the newly emerging molecular biology. The place of transfer RNA was expressed in the language of molecular biology as a stage in the process of information transfer from genes to proteins.[3]

To draw attention to what he considers to have been marginalized in previous accounts of experimental science, Rheinberger has self-consciously subordinated the motivations, intentions, and thoughts of the individual scientists making up Zamecnik's group to the performance of the experimental system in whose development they shared. Howard Gruber has argued, on the other hand, that it is necessary to reconcile the crucial role that is attributed to chance in the creation of knowledge (within which category we might provisionally include the unforeseen directions in which an experimental system may lead its creators) with the fact the "creativity is *purposeful* work." When a person is "'purposeful,' we mean that he or she cannot easily be deflected from the pursuit of a chosen course. Together, the deflections and the responses illuminate the purposes of . . . the striving creative subjects themselves." If it is agreed that "chance and purpose both play a role . . . the key point is that the evolution of human purpose transforms the operation of chance."[4] Applying Gruber's perspective to Rheinberger's analysis, we might wish to restore to the foreground in the treatment of experimental systems what Rheinberger has displaced to the periphery. Investigators can choose whether to follow or not to follow the unforeseen directions opened up by the capacities of their system. If they allow the system to carry them where they had not previously intended to go, one should regard them not simply as having become captive to what they have created, but as making a decision that the new direction fits within the broader purposes of their scientific enterprise. We can find, in fact, other case histories in which investigators have abandoned experimental systems on which they had long relied because their purposes directed them instead to invent or adopt different systems whose capacities enabled them to maintain the direction of the investigative pathway they had previously been following.[5]

Finally, we might comment that the unpredictability in scientific research that Rheinberger has located within the evolving capacities of experimental systems represents only one class among the many types of event that can divert investigators from their prior intentions and set them off in unforeseen directions. In the experiences of the subjects of my detailed studies we shall encounter some of the other classes of events that impinge in such ways on investigative pathways.

The experiments of Antoine Lavoisier often turned out in ways that he had not expected, but these outcomes usually indicated to him that they were flawed,

and he typically sought to modify them, or to invoke assumptions about experimental errors, until he had attained what he intended. From the time he took
up his agenda in February 1773, to study the processes that absorb or release air,
his overriding purpose, to construct a theoretical framework within which he
could unify these processes, dominated his investigative course. He allowed neither recalcitrant difficulties with his experimental systems, nor unexpected observations, to divert him from that goal. Nevertheless, at least two unpredicted
events were critical to the shape and outcome of his endeavor.

Three experiments that Lavoisier had performed in the fall of 1772 provided
the starting point for the extended investigation he began the following spring.
He found that phosphorus and sulfur burned in a closed container gained
weight and absorbed air, and that a calx of lead reduced in a closed space in the
presence of charcoal lost an amount of weight equal to the weight of air that it
gave off. Although these experiments have the appearance of being designed to
demonstrate what he had already expected to happen, Lavoisier wrote in a
manuscript that Carl Perrin has dated to September 1772—that is, very close to
the time in which he carried them out—that he had been led to do them "more
by accident than by theory."[6] Too little documentation for this period has survived to allow us to reconstruct either the nature of the accident or Lavoisier's
expectations prior to observing the results of the experiments. The statement itself raises the intriguing possibility, however, that it was an event that he had not
predicted that initiated the entire project Lavoisier pursued for the next two
decades with such momentous consequences.

When he carried out these first experiments, Lavoisier assumed that the "air"
involved in each of these processes was simply the ordinary air of the atmosphere,
generally regarded at the time as a simple, unitary substance. By the time he took
up the investigation in February, however, he had immersed himself in the literature, from Stephen Hales to his own time, on the subject of airs, and had
come to realize that the identity of the air, or "elastic fluid," that took part in the
combustions and metallic reduction he had studied was the central question still
to be resolved. He at once raised several possibilities: that it was the air itself entire, the air combined with some volatile part of the body from which it em-
anates, or a substance "extracted from the air of the atmosphere." Finding in several of his early experiments that metals heated in a closed space were calcined
only to a limited degree, he came to suspect that the "totality of the air we respire
does not enter into the metals that one calcines, but only a portion which is not
present in abundance in a given mass of air."[7]

The reduction of metallic calxes, which Lavoisier carried out in the traditional way mixed with charcoal, yielded an air with the special characteristics of the distinct species that Joseph Black had first identified and named "fixed air." That is, it was very soluble in water and formed a precipitate with a solution of lime-water. By the time he presented his first paper on the subject to the Académie des Sciences in April 1773, Lavoisier had fastened onto the theory that fixed air was the substance exchanged in each of the processes he studied, that phosphorus and sulfur absorbed fixed air when burned, and that metallic calces consisted of the metal joined to fixed air.[8]

By the end of the summer of 1773 Lavoisier had extended his experiments beyond the preliminary and rather indecisive ones on which he had based the theoretical structure that identified fixed air as the substance involved in each of these processes. Instead of confirming this conclusion, however, his later results made the identification less certain. In the papers that he presented to the Academy intended to lay out the details of the experimental evidence for his theory, he no longer referred to the air as fixed air, but generically as the "elastic fluid" absorbed in combustion and calcinations and released in metallic reductions.[9]

His inability firmly to identify the elastic fluid in question did not induce Lavoisier, however, to doubt that there was only one such substance involved in all these processes, and he persisted in the belief that it could eventually be shown to be fixed air. When he undertook to reduce the calx of mercury without charcoal early in 1775, he still apparently expected that he would obtain fixed air, just as in the reductions of other metals with charcoal. His expectation may have been reinforced by the fact that another Parisian chemist, Pierre Bayen, who preceded him in carrying out this operation, identified the air released as fixed air. When Lavoisier subjected the product he obtained to the standard limewater test, however, it did not form a precipitate, only turned the solution slightly opaline. More striking still was that when he placed a candle in it, the flame was not extinguished, as it would be in fixed air. Moreover, when he applied the nitrous air test devised by Joseph Priestley to the air, it contracted in volume to about the same degree that ordinary air did. These results forced Lavoisier to conclude that the air was "in the state of common air," except that "it retains a little of the nature of inflammable air." When he presented these results to the Academy in April 1775, he acknowledged that he had recognized the negative reaction of the air obtained to each of the tests for fixed air "with great surprise."[10]

This outcome, unpredictable within the theoretical framework that had guided Lavoisier for the previous two years, changed the nature of his investigation. It was now clear that no single air could fulfill the role he had assigned to an "elastic fluid," but that at least two airs were involved—fixed air and an air which seemed at first to be common air, but which he noticed soon afterward was by the usual tests "better" and "purer" than ordinary air. The result that so surprised him set two new problems for Lavoisier. The better-known one was the identity of this air itself. As has often been told, it was Priestley who first realized that the air was sufficiently distinct from common air so that he described it as a new species, which he named "dephlogisticated air," and Lavoisier who identified the air as one of the two main components of the atmosphere that he named, eventually, oxygen. An equally pressing problem for him, however, was the relation between this air and fixed air. If there were two airs involved in processes in which he had thought there was only one, then there must be some process through which one air is converted to the other. He reached the solution to this problem only in 1777.[11]

Was the surprising result that complicated Lavoisier's theoretical problem and changed the course of his further investigation in 1775 generated by the capacities of the experimental system with which he was operating? That he had devised such a system, an assembly of apparatus and procedures that enabled him to measure the volume of an air absorbed or released in the processes that interested him, together with several identification tests for the qualitative nature of the airs involved, and that this system became more capable and more widely applicable to the extent that he gained experience with it, underlay much of his success. It was not the realization of new intrinsic capacities of his system, however, that led him to the experiment on the reduction of mercury calx without charcoal. Rather, his theoretical reasoning prompted him to try the experiment, because he was concerned that the presence of charcoal complicated the interpretation of metallic reductions. That is, he could not be certain whether the fixed air he had received in his previous reductions derived from the metallic calx or the charcoal, and it was to remove that latter possible source that he established the experimental conditions that led to this unpredicted but critical outcome.

We may note again that the unexpected discovery that forced Lavoisier to abandon his belief in a single air underlying all the phenomena he was studying did not divert him from his central purpose of constructing a unified theory of combustion, calcination, and metallic reductions; rather, it changed the nature of the problems he had to solve in order to achieve that goal. With Claude Bernard, on the other hand, unexpected observations frequently altered his goals, leading him to discoveries he had not initially set out to make. Such unpredicted events happened to him so regularly that he raised them to a general feature of scientific reasoning. "Scientific investigations and experimental ideas," he wrote in *An Introduction to the Study of Experimental Medicine,* "may have their birth in almost involuntary chance observations which present themselves either spontaneously or in an experiment made with a different purpose."[12]

The examples with which he illustrated this generalization show that such unexpected events played a part in each of Bernard's major discoveries, as well as in many of his minor ones. The accident that rabbits brought into his laboratory excreted clear, acidic urine because they had been fasting was the starting point for experiments that led him to characterize the differences between herbivorous and carnivorous urine. The chance observation that the lacteal vessels of rabbits fed on meat became milky only thirty centimeters beyond the pylorus, whereas he remembered that in dogs he had seen milky lacteals much closer to the pylorus, led him to reflect that the difference must be related to the location of the opening of the pancreatic duct in rabbits at that point in their small intestines. Reflecting that pancreatic juice might, therefore, be responsible for the digestive emulsion of fats, he devised the method to procure pancreatic juice through a canula that enabled him to verify this hypothesis.[13]

We have already seen that Bernard had been seeking to locate the place in animals at which sugar ingested in the food disappeared, when the result of a control experiment with an animal fed on meat diverted him instead to the question of where sugar is secreted in an animal not receiving it in its diet. While seeking in 1855 to determine the amount of sugar that can be obtained from the liver by flushing out its blood, and intending to make two duplicate analyses, Bernard made one analysis just after killing the animal, but did not have time to perform the second one until the next morning. To his surprise, he obtained much more sugar the second time than the first time. Reflecting on the cause of this unpredicted variation led him to infer that the liver continues to produce sugar for some time after its "death," and consequently to conclude that the source of the sugar is not in the circulating blood, but an insoluble substance deposited in the liver tissue.[14]

When Bernard severed the cervical sympathetic nerve on one side of a rabbit in 1851, he expected that the resulting paralysis would slow down combustion in the blood and cause the affected parts to be cooled. "What happened," he related, "was just the reverse." The circulation became more turgid, and the temperature increased on the affected side. Accordingly, he "at once abandoned theories and hypothesis to observe and study the fact itself," a course which led him eventually to "open up a new path" for the investigation of thermo-regulation by means of the vaso-motor system of nerves.[15]

In those examples for which Bernard's investigative pathway has been reconstructed in greater detail from his laboratory notebooks, we can see that he to some extent idealized the actual progression of events, simplifying them in ways that cast into sharper relief the features of scientific reasoning that he wished to stress. Nearly two years passed, for example, between the time that Bernard recalled observing the milky lacteal vessels at a distance from the pylorus in rabbits, and the time that he first obtained pancreatic juice through a canula and observed its action on fats. Had he been strongly impressed in 1846 with the idea that pancreatic juice may have a special action on fats, he would have been able to test his hypothesis much sooner than he did. Long before he succeeded with the operation to procure the juice from its duct, he had been using extracts of pancreas tissue to test the action of pancreatic juice on meat and carbohydrates, but never included fats in these experiments. Even after the canula operation, he applied the freshly obtained juice to a long series of tests on meat and starch before turning to a single test on candle tallow. In his laboratory notebook he wrote the following description of that result: "After 8 [hours] of continued digestion (the liquid is very plainly alkaline), a white, perfectly homogeneous emulsion has formed. The liquid is fine, like milk. . . . There has been, therefore, a peculiar action of the pancreatic juice on the fat. . . . It will be necessary to make other comparative experiments on the subject."[16] If we accept Bernard's own retrospective account, this result confirmed the hypothesis he had drawn two years earlier from his chance observation of the rabbit lacteal vessels. If, on the other hand, we ask whether the lapse of time and the low priority he seemed to place on this test on the day when he finally performed it suggests that he had no strong expectations about the outcome, then this result itself may constitute the unpredicted event that soon became Bernard's first major discovery.

In his account of the effort to trace the disappearance of sugar that led instead to the discovery of its source in animals, Bernard wrote:

> I conceived the hypothesis that sugar introduced into the blood through nutrition might be destroyed in the lungs or in the general capillaries. The theory, indeed, which then prevailed, and which was naturally my proper starting point, assumed that the sugar present in animals came exclusively from foods, and that it was destroyed in animal organisms by the phenomena of combustion, i.e. of respiration. . . . But I was immediately led to see that the theory about the origin of sugar in animals, which served me as a starting point, was false. As a result of the experiments . . . I was not indeed led to find an organ for destroying sugar, but, on the contrary, I discovered an organ for making it, and I found that all animal blood contains sugar even when they do not eat it. So I noted a new fact, unforeseen in theory, which men had not noticed,

doubtless because they were under the influence of contrary theories that they had too confidently accepted. I therefore abandoned my hypothesis on the spot, so as to pursue the unexpected result.[17]

Bernard's laboratory notebook preserves the record of an experiment, carried out in early August 1848, that is very likely to be the one whose result was contrary to his expectations. After feeding a dog on raw meat for eight days, he killed it and extracted blood from its portal vein, heart, and a wound in the neck, as well as lymph from the thoracic duct. The blood from the portal vein produced an "enormous reduction" with the reagent on which Bernard relied to identify sugar, whereas that from the heart yielded a clear but less abundant reduction. "How did it happen therefore," Bernard asked, that "there was sugar (or a material that reduces) in the blood of the portal vein?" Because that vessel absorbs material directly from the alimentary tract, Bernard searched the stomach and intestines, but found no sugar there. "This experiment is exceedingly strange," he then wrote down:

> From it one can comprehend nothing. Would sugar form in the portal vein, by what organ, by what mechanism?
>
>   It will be necessary to take this blood from the portal vein of a dog in abstinence and see if one will find there that material which reduces.
>
> . . . . .
>
>   That reducing material . . . disappears quite rapidly, for the blood of the heart contained less of it and the blood from the neck only in a very equivocal fashion. What therefore is the organ that would form that sugar or that reducing material?[18]

The experiment and Bernard's response seem to fit his later account so closely, that Mirko Grmek was led to treat it as confirmation of the story Bernard told about the hypothesis that had formed his starting point and his abandonment of that hypothesis in the face of a "new fact" contrary to it.[19] The broader narrative of the history of Bernard's investigations over the preceding five years, however, raises some doubts. The "prevailing theory" was that maintained in a series of lectures delivered by the leading French chemist Jean-Baptiste Dumas in 1841, and subsequently published under the joint names of Dumas and his collaborator, Jean-Baptiste Boussingault. The theory that animals receive all of their bodily constituents ready-formed in their nutrients and only decompose them by stages by combustion processes had a large local impact, but was immediately challenged by others, in particular by the German chemist Justus Liebig, who maintained that animals can convert carbohydrates to fats. By 1845 the French chemists who attempted to defend their viewpoint by feeding experiments had instead proven Liebig right, so that Dumas's assertion that animals consume only what they receive from plants no longer held in its full generality by 1848. Moreover, as a student of the skeptical Magendie, Bernard had from the outset of his career questioned the adequacy of theories derived from chemical considerations alone, without being tested by experiments on living animals. Consequently, although it may well have been that Bernard began his search for the disappearance of sugar assuming that its origin was nutritive, it is unlikely that he adhered so strongly to this purely chemical theory that an experiment seeming to contradict it would have aroused the bewilderment expressed in his notebook comments. One would, instead, have expected him immediately to welcome this further sign of the insufficiency of the theories of the chemists. My own interpretation of Bernard's comments is that what baffled him was not so much finding sugar in the blood of an animal not nourished on sugar, but finding it in abundance specifically in the portal vein, because he could not imagine what organ could produce it there. This contradiction was removed only when he subsequently was able to show that the liver secretes sugar into the blood, and that its abundance in the portal vein was due to a reflux of blood in the opposite direction from its normal flow, caused by the release of pressure in the portal vein when he dissected the abdomen of the animal.[20]

If the notebook record complicates the story that Bernard told, it nevertheless supports the view that his most important single discovery was the outcome of an unpredicted event. In its aftermath Bernard *did* abruptly change the direction of his investigation, giving up his effort to identify the place where sugar is destroyed (an effort doomed from the start to failure, because his whole conception that the sugar disappears somewhere during its circuit in the blood was wrong), and directing his experiments intensely toward the location of the nonnutritional source of the sugar he had found.

Bernard did not employ for most of his work an experimental system in the literal sense in which Rheinberger has described such systems as a ubiquitous feature of twentieth-century biological research. His experiments relied mainly on his extraordinary skill with the scalpel, and the rather more limited range of chemical methods that he learned to apply. One might, however, draw an analogy between the growing repertoire of vivisection operations that he mastered during the course of his investigations and Rheinberger's conception of the increasing realization of the capacities of an experimental system. It is, I believe, not coincidental that the way to both of Bernard's first two major discoveries was paved by his learning to perform a novel operative procedure that required the best in his remarkable surgical prowess. He succeeded in the delicate operation of procuring pancreatic fluid through a canula inserted in the duct so as to direct the flow outside the animal only after numerous attempts in which the animal had died too soon for him to collect the fluid. His detection of sugar in various parts of the circulation during his search first for the site of its disappearance and then of its source required similarly refined skills to withdraw samples of blood from deep within the circulatory system. Once he learned to carry out such specialized and demanding operations readily, he was able to use relatively simple chemical procedures to identify, respectively, the action of pancreatic fluid on fats, and the presence or absence of sugar in the blood.

We might say, then, that something corresponding to the growing realization of the intrinsic capacities of an experimental system led Bernard to at least some of his unforeseen discoveries. Because the surgical skills and procedures he accumulated were not systematically assembled in the way the components of an apparatus are, however, Bernard may have retained greater versatility in applying them and devising others in different combinations to pursue different problems and, unlike Zamecnik's group, adding new dimensions within his investigative pathway without abandoning older ones.

At least one unforeseen event was essential to the pathway leading to each of the significant discoveries that Hans Krebs published between the time he began his own research in 1930 and the presentation of the citric acid cycle in 1937. Some of these unexpected turns were generated internally in the course of his investigations, while others came to his attention from contemporary work carried out in other laboratories.

The dramatic increase in the rate at which a liver tissue slice produced urea when Henseleit added ornithine together with ammonia to its medium was, according to Krebs's subsequent discussion of the result, entirely unexpected. Prior to ornithine, they had tested four or five common amino acids together with ammonia and had observed no appreciable increase over the rate with ammonia alone. Nor would the standard view of urea synthesis have given them reason to suppose that any single amino acid would stand out from the others. The mechanism was supposed to be the same for all amino acids, the first step being the deamination of the amino group common to them.

If he had no reason to expect that ornithine would have this effect, why did Krebs decide to try it? His own recollection that he was systematically testing all amino acids does not fit with the notebook record showing he had used only several ordinary amino acids before trying this rare and expensive one. Thinking that he must have had some idea, since forgotten, to lead him in this direction, I proposed a possibility, but it did not fit his memory and he did not agree with the suggestion. Whether he had had some such reason, or just came across it by some chance circumstance, the fact that in combination with ammonia, ornithine nearly doubled the rates he had previously observed completely changed the direction of his investigation. Up until then he had mainly tested previously published theories, and his project looked as though it might merely confirm in his more precise experimental system what had already been believed about urea synthesis. After he observed the ornithine effect, the whole thrust of his investigation turned to the effort to explain it, and his ability to do so culminated in his first major achievement.

In the spring of 1933, Krebs was simultaneously pursuing his recently initiated studies of the processes through which foodstuffs are decomposed and the deamination study he had begun the previous year, often alternating almost daily between experiments on the one and the other problem. In his deamination experiments he had been able to establish for several amino acids, by measuring the consumption of oxygen, the quantity of the corresponding keto acid produced, and the increase in the amount of ammonia present, that these quantities were in the ratios expected according to the assumed equation for the oxidative deamination reaction: (2 amino acid $+O_2 = 2NH_3 + 2$ keto acid).[21]

At the end of March Krebs fixed his attention on two amino acids, aspartic and glutamic acid, that he had found to have particularly high rates of deamination in earlier experiments. Both of them happened to be dicarboxylic acids. He had so far been unable to isolate the keto acids that should be formed, because, as he thought, they were further decomposed. Having been able in some previous investigations to block the decomposition of one of them, alpha-keto glutaric acid, by adding arsenite to the medium, he tried the same reagent in an experiment on the deamination of aspartic acid, and found that the rate of ammonia formation was increased by its presence. This was a puzzling result, because blocking the decomposition of the product ought not to have altered the rate of the deamination reaction itself. When he tried the same procedure with glutamic acid, he encountered another anomaly. Comparing the quantity of ketoglutaric acid formed to that of ammonia, he observed "more keto acid than NH_3." From the stoichiometry of the reaction these quantities should have been the same, as he had already found in the case of other amino acids. To check this anomaly he next performed comparative experiments on glutamic acid and lysine and found that the arsenite increased the amount of NH_3 formed very little with the latter, but enormously with the former.[22]

Shortly after carrying out these experiments Krebs was barred from his laboratory in Freiburg in accordance with the Nazi legislation dismissing non-

Aryans from the German civil service. During his enforced leave he wrote several papers on the work he had done up until then, including a short note on the surprising results of the experiments on glutamic and aspartic acid. He interpreted the fact the arsenite increased the amount of ammonia present afterward to the fact that it must block the formation of some nitrogen compound that otherwise consumed some of the ammonia. "The question, which nitrogen compounds arise from ammonia . . . and the amino acids, especially from glutamic acid in the unpoisoned kidney cell," he ended the paper, "I could not answer, since I had to break off the work."[23]

As we have already seen in the previous chapter, Krebs took up the question what became of the nitrogen released in the deamination of glutamic and aspartic acid again in Cambridge in August 1934, an investigation that led to his discovery of the synthesis of glutamine. The behavior of ammonia in the deamination of glutamic acid that had begun as a surprising anomaly now became an expected feature of the reaction by which glutamic acid is converted to glutamine. Once when I asked Krebs what role chance had played in his discoveries, he picked out this episode as the most prominent example. He had, he said, been looking for the formation of NH_3 but by "serendipity" he found instead its disappearance.[24]

In the investigation that led immediately to the discovery of the citric acid cycle, the most significant unforeseen event was the appearance of the paper by Carl Martius and Franz Knoop that gave Krebs the critical clue concerning the likely pathway of decomposition of that metabolite. It has often been assumed by biochemists, and Krebs himself sometimes hinted, that he had all along been looking for a second cycle similar in nature to the ornithine cycle of urea synthesis that he had discovered five years earlier. The detailed reconstruction of his unbroken pathway during the intervening years reveals, however, that until he read this paper, Krebs had not been seeking a cyclic solution to the problem of oxidative carbohydrate breakdown. Far from it, he was preoccupied for more than a year with finding linear sequences of dismutation reactions to account for that process. It is most probable that he came across Martius and Knoop's paper, at the earliest possible moment of its appearance in the Sheffield library, because he was in the habit of using his spare time to scan through the latest issues of biochemical journals. It was thus this fortunate chance encounter that changed the direction of his search and led quickly to the success that had eluded him in his earlier efforts.

I have focused here on three unpredicted events, each of which proved essential to one of Krebs's important discoveries. In his experimentation, however, unexpected results were far more frequent, and they often caused him to alter course in an effort to follow up a lead they suggested. In most cases, however, these diversions lasted for only a short while. Either subsequent experiments failed to support what the first ones suggested, or he decided not to be distracted further from his main line of investigation.

The Warburg manometric tissue slice methods on which Krebs relied for most of his investigations during the first decade of his career constituted a well-defined experimental system. As the above examples described and the many smaller ones I could mention indicate, this system comprised a very efficient engine for generating surprises. Over the short term these surprises often diverted Krebs toward unforeseen new directions. Over the longer term, however, he most often returned to his previous goals. "I have to restrain myself," he once commented, "not to scatter my interests, and to complete a piece of work."[25] His ability both to be easily diverted and to complete what he had previously set out to do was facilitated by the particular characteristics of his experimental system (Figure 13). The versatility of the manometric methods he applied, as well as the ease with which one could carry out two complete sets of multiple experiments in one day, enabled him to follow many unexpected leads far enough to see whether they would prove fruitful, without leaving the threads of his previous projects behind long enough for the trail to grow cold. Here chance and sustained individual purpose interacted in ways that sacrificed neither to the imperatives of the other.

To remain productive, Rheinberger asserts, an experimental system must be sufficiently open to generate unprecedented events by incorporating new techniques and devices, but sufficiently closed to maintain its reproductive coherence. "It has to be kept at the borderline of its breakdown." He invokes the "principle of limited sloppiness" commonly attributed to Max Delbrück. According to this principle, if the experimenter is too sloppy he will not attain reproducible results, but a little bit of sloppiness allows unexpected phenomena to appear.[26] The context in which Delbrück had made this remark, at Oak Ridge in 1949, was the discovery by Renato Dulbecco of the photo-reactivation of bacteriophage. After subjecting phage to inactivation with ultraviolet radiation, he infected bacteria with them, preparing duplicate plates of those that had received the same dose. He then noticed that the plates among the duplicates placed nearer to an ordinary light bulb produced a greater number of plaques than those further from it. The "limited sloppiness" that induced this discovery implies

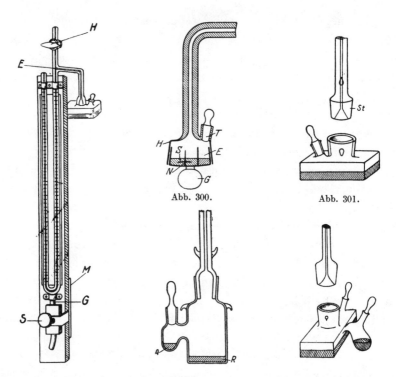

Warburg manometer of the type used by Hans Krebs for experiments in intermediary metabolism. Left, overall apparatus. Tissue slices surviving in the physiological salt solution are placed in a cup suspended at the right side of the apparatus. Thumb screw is adjusted to maintain level of fluid constant in left-hand, open tube. Differences in height between fluid in two sides over the course of the experiment are read as pressure differential and converted to volumes of gas consumed or emitted by the tissue. Middle and right side: variations on the configuration of the cups containing tissue slices. Side bulbs are used for tipping in metabolites or blocking agents during the experiment.

that if Dulbecco had controlled the conditions more rigorously, so that all plates received the same illumination, he would not have observed the effect that led him to the phenomenon of photo-reactivation.[27]

Undoubtedly unforeseen events that open up new investigative pathways or discoveries do result sometimes from limited sloppiness, or from experimental systems on the edge of breakdown; but the example drawn above from Krebs's work shows that they can emerge equally well from the precise control afforded by an experimental system. It was because his methods allowed him to measure the ammonia and keto acids produced and the oxygen consumed in his deamination experiments with sufficient accuracy and reliability to expect them to confirm the stoichiometric ratios of the presumed reaction, that he was able to treat a departure from the expected ratio not merely as experimental error, but as an indication that some of the ammonia produced was disappearing in another, unknown reaction.

At Cold Spring Harbor during the summer of 1941, Salvador Luria and Max Delbrück infected a strain of the bacterium *E. coli* simultaneously with two different bacteriophages. During the preceding three years Delbrück had intensively studied the reproductive cycle of bacteriophage, focusing on the "one-step growth curve": that is, the process by which the phage are adsorbed onto bacteria, undergo a latent period, then cause the bacteria to lyse, releasing progeny phage into the medium. Though he had been able to describe quantitatively various relations between the "input" and "output" of the process, he was unable to discern through these methods what happens in between, how the phage actually reproduce within the bacterial host. As he and Luria explained in their published report on their experiments with the simultaneous infection of bacteria with two viruses, "The growth of a bacterial virus . . . occurring only in the bacterial cell, may be said to proceed behind a closed door. The experimenter can follow the virus up to the moment when it enters the cell, and again after liberation from the cell. There is, as yet, no way of telling what goes on within the cell." They hoped that one of the two viruses with which they infected the cells might lyse the bacteria while the other was still growing. "Thus an intermediate state of virus growth might be revealed." They tested the two viruses under various conditions, but their "expectation did not materialize." The unpredicted result of their experiments was that mixed infections always suppressed the growth of one virus completely, while the other grew normally. The experiments failed in their original purpose, but revealed instead an unanticipated phenomenon, which they named "interference" (later "mutual exclusion"), and which Delbrück subsequently studied for several years.[28]

Meselson and Stahl encountered many unpredicted events during the course of the investigation from which their notable experiment emerged. Some of these occasions only raised obstacles or diverted them temporarily from their primary goal, to solve by means of density differences the much-debated problem of DNA replication. One of these events, however, became essential to their success in that quest.

When they first devised plans to distinguish, by means of the resulting density differences, normal DNA from DNA into which they would substitute an isotope or a base heavier than the ordinary four bases of the double helix, Meselson and Stahl had in mind that they would choose a medium for the centrifuge tube whose density was intermediate between that of the "heavier" substituted DNA and the ordinary molecules. The lighter DNA they thought would collect at the top of the centrifuge tube, from which they could then remove it, while the heavier DNA would sink to the bottom, where it could be extracted through a hole punched afterward in the plastic centrifuge tube. If, as they expected, there would be produced during the replication also "hybrid" DNA of intermediate density, they would separate it from either the light or heavy material through further centrifugations in media of appropriately chosen density. After a systematic examination of the densities given for various salt solutions in handbooks of physics and chemistry, Meselson concluded that the medium most likely to meet their needs was cesium chloride.[29]

In their first experiments, in September 1956, in which they incorporated the base 5-bromouracil into the DNA of bacteriophage T4, they used a "preparatory" ultracentrifuge. Preparing a cesium chloride solution denser than that of the phage with unsubstituted DNA that they placed in it, they were able to collect the phage afterward from the surface of the solution. Before they turned to a solution of expected intermediate density, however, Meselson decided to carry out the operation in an analytical ultracentrifuge, whose optical system would allow him to detect density gradients in the cell, as well as the location of molecules, such as DNA, that absorb ultraviolet light, during the course of the centrifuge run (Figure 14).[30]

The optical system through which one could monitor density gradients formed along the axis of the centrifuge cell was known as a Schlieren line. If this line remained horizontal along its base line, the density was uniform through the cell. A rising line indicated a density gradient at whatever point along the axis, represented by the horizontal dimension, it occurred. If the line as a whole moved vertically, the gradient would be interpreted as uniform, whereas sharp peaks would indicate places along the axis where density discontinuities occurred. In his first run, using whole phage with normal DNA, Meselson chose a cesium chloride solution of density just enough greater than the known density of the phage so that he would expect to float the phage to the top. Watching the Schlieren line, he probably did notice a peak that moved upward in the cell, but his attention was caught by the fact the line as a whole began rising, and kept rising slowly throughout the run. This observation "astounded" him. He knew from the literature that small molecules such as those of inorganic salt solutions did form density gradients in a centrifuge at equilibrium, but because the force moving them decreased their molecular weight in proportion, he ex-

Matthew Meselson operating the Model E Analytical Ultracentrifuge. Photo courtesy of the Archives, California Institute of Technology.

pected that it would require many days for such a solution to come to equilibrium. It took him at least a day to accept the reality of the cesium chloride gradient that he had observed.[31]

This unexpected turn did not immediately cause Meselson and Stahl to rejoice (Figure 15). They realized that, if the density difference between the salt solution at the top and the bottom of the cell were greater than the density difference between substituted and unsubstituted DNA, then these molecules would not collect, respectively at the bottom and top of the cell, but somewhere in between, where their buoyant density equaled that of the medium. If they were lucky, bands would form that they would be able to see by means of the ultraviolet absorption. They did not know, however, how wide the density spread of the cesium chloride solution would be. The bands would be expected to have a certain width dependent on the equilibrium between centrifugal forces driving the molecules toward the level in the medium equal to their density, and diffusion tending to spread them out. If the width of the bands formed turned out to be greater than their separation due to the differences in the density of substituted and unsubstituted molecules, this would defeat their purposes in carrying out the experiments.[32]

It required two months of experimentation for Meselson and Stahl to establish that the density gradient in a cesium chloride solution happened to be just

Franklin W. Stahl at Cold Spring Harbor in 1958. Photo
courtesy of Cold Spring Harbor Archives.

large enough to separate 5-bromouracil substituted phage DNA widely from unsubstituted DNA. For other reasons, however, their attempts to perform the "transfer" experiment on phage and with 5-bromouracil never yielded satisfactory results. When they turned, in the fall of 1957, to ^{15}N substituted DNA in bacteria, however, the very first, somewhat flawed experiment showed that their system would work, and their second attempt produced the beautifully clean-cut result that has made the Meselson-Stahl experiment classic. An unpredicted event that had first appeared as a potential obstacle turned out instead to be the foundation for their brilliant success.[33]

The unexpected observation of a cesium chloride density gradient transformed Meselson and Stahl's project to solve the DNA replication problem shortly after they had taken it up. For Seymour Benzer it was an unpredicted and equally surprising event that induced him to begin the project that led to the fine-structure mapping of the rII region of bacteriophage T4 (Figure 16). At the beginning of January 1954, Benzer had plated out some T2r mutants and T2r$^+$ ("r" standing for a class of mutants resulting in rapid lysis of the bacteria in which the phage replicate, "r$^+$" indicating wild type) on a strain of the bacterium $E.\ coli$ that he happened to have in his laboratory. Coincidentally he was preparing at the same time to demonstrate the phenomenon of lysogeny to the phage course he was teaching. Lysogenic bacterial strains are those in which an infecting phage may remain incorporated for an indefinite number of generations without lysing the bacteria. The lysogenic strain was designated K12(λ). Benzer plated the mutant and wild type T2r strains both on the lysogenic bacterial strain and on a nonlysogenic strain that had been derived from it. On the latter, both strains formed small, fuzzy plaques, but on the K12(λ) plate to which he had added the mutant strain, he found no plaques at all. This result so surprised him that he thought at first that he had forgot to add the phage, but he repeated the experiment and got the same result.[34]

Now the significance of this result seemed obvious to him. It offered an unprecedented opportunity to detect mutants along the phage genome with a much higher degree of resolution than had been obtained in the genetic mapping of other organisms. If r mutants would not grow on the lysogenic bacterial strain, then when he prepared stocks containing two such mutations, only wild types resulting from a recombination event would produce plaques. He could, in this way, detect such events even if they occurred at the extremely low frequencies that would be expected of mutations very close to one another on the genome. The prospect was so inviting to him that he soon abandoned the projects on which he had previously been engaged, and was able to show by early

Page from looseleaf laboratory notebook of Seymour Benzer, dated May 10, 1954. The experiment consists of crosses between three T4r mutants, and includes the earliest sketch by Benzer of a genetic map for these mutants. By 1957, Benzer had mapped hundreds of r mutants in this region.

summer of that year that he would be able eventually to press the resolution of his system down to the dimensions of a few DNA nucleotides.

Heretofore in this chapter I have treated chance events as synonymous with unpredicted, or unexpected ones. A comparison of the experiences of Benzer on the one hand, and of Delbrück and Luria and of Meselson and Stahl on the other, shows, however, that chance events form only a class within the broader category of unpredicted ones. Had Benzer not happened to plate a particular mutant strain of bacteriophage on a particular lysogenic bacterial strain that he had around for an unrelated purpose, he might not have observed the phenomenon that redirected his investigative pathway. Once Delbrück and Luria chose to infect bacteria with two kinds of phage at once, however, it became inevitable that they would encounter the interference that blocked their original intentions but opened up a new opportunity for further exploration. Similarly, when Meselson and Stahl chose to separate DNA of differing densities in an analytical ultracentrifuge with cesium chloride as the medium, they were bound to encounter the density gradient that redirected their investigative pathway. Though different in their underlying nature, both chance events and inevitable but unexpected ones contribute to the high degree of unpredictability that characterizes the research enterprise.

Yet, though unpredictability is, as Jacob expressed it, so central to the "very nature of the scientific enterprise" that "one can never say how it will evolve," the enterprise cannot be thoroughly unpredictable. Otherwise it would be impossible to proceed rationally or productively. If there is "no way to tell where a given domain of research will lead," investigators must, nevertheless, have reasonable confidence that their pathways will lead them to something significant enough to reward the great investment of effort they must devote to their work. Otherwise the scientific enterprise could not have flourished as it has done for the past three centuries.

Despite the diversions and other unexpected events that each of my subjects encountered along the way, each of them achieved in the long run much of what he set out to do. They could not foresee in detail how things would turn out, but they were able to respond to the unforeseen in ways that fulfilled their broader goals. Although my examples are not entirely representative, since the scientists I studied were selected in part because of their conspicuous successes, the experiences they met along the way do suggest some of the reasons that the unpredictability of research is not unlimited, that the unexpected occurs within certain boundaries set by the more general rules within which a domain of science is practiced. When Lavoisier realized that there were two elastic fluids implicated in the processes he was studying, where he had long sought to identify a single one, that fact confronted him with a complication, but not with an unfathomable situation. The "pneumatic chemistry" of his time had already produced evidence that there were numerous species of airs, and he had already come to suspect that the atmosphere itself contained more than one species. His unexpected result could, therefore, be fitted within an existing general conceptual framework.

Similarly, neither his unexpected discovery of the unique action of pancreatic juice on fat, nor that of the production of sugar in the liver confronted Claude Bernard with phenomena to which current views of nutrition could not readily be adjusted. Others had already been looking for the agent responsible for the digestion of fats, and Bernard's result provided an unexpected resolution to a well-defined problem whose general premises it did not violate. Although his discovery that animals produce sugar independently of their nutrition did conflict with the particular grand theory proposed by Dumas, that theory itself was not deeply embedded in the structure of the field, and the discovery did not conflict with any more firmly established contemporary physiological or chemical knowledge.

Each of the metabolic pathways Hans Krebs discovered during the 1930s was in its specific form unpredicted, but none of them violated any of the more general criteria that biochemists had laid out for investigation of these processes. Each gave unexpected forms of specificity that conformed to accepted general principles. Neither the unexpected rapid formation of a cesium chloride gradient that Meselson observed, nor the inability of T2r mutants to grow on a particular lysogenic bacterial strain that Benzer observed overturned general principles guiding research in the relevant scientific domains. Once observed, the unpredicted easily became the accepted foundation for further progress.

In their book *Laboratory Life,* Bruno Latour and Steve Woolgar asserted that research is the effort to bring pockets of order out of disorder. "Each scientist strives to get by," they believe, "amid a wealth of chaotic events." A massive background of noise continually threatens to drown the signals they seek.[35] This is, I believe, a misconception of the research process. Unpredicted events are not necessarily chaotic events. Investigators operate within highly ordered frameworks shaped by all the past work in their domains, and events can be recognized as unpredicted only by their deviation from that which the preexisting order does predict. The unpredicted events to which the pursuit of investigative pathways so regularly leads researchers create temporary pockets of disorder, and successful responses to these events most often involve adjusting the previous order at its edges sufficiently to fit the seemingly disorderly result into its texture.

Readers may notice some resemblance between the preceding statement and Thomas Kuhn's conception of normal science as research carried out under the guidance of a prevailing paradigm. In a general sense that is true, but my position does not entail acceptance of the entire structure of normal and revolutionary science proposed in *The Structure of Scientific Revolutions*. Kuhn's dichotomy represents, I believe, two ends of a broad spectrum, between which the research of leading investigators often has an intermediate character. Not only the means by which a puzzle is solved, but the nature of the solution is often more surprising and more original than his characterization of such work as "mopping up exercises" would imply.

Are scientific discoveries always the outcomes in part of unpredictable events? In his *Structure,* Kuhn wrote, "Discovery commences with the awareness of anomaly, i.e. with the recognition that nature has somehow violated the paradigm-induced expectations that govern normal science. It then continues with a more or less extended exploration of the area of the anomaly. And it closes only when the paradigm theory has been adjusted so that the anomalous has become the expected."[36] Most research within normal science, which does not encounter such novelties, does not produce discoveries, but only the solutions to puzzles set by the paradigm itself. The investigator does not seek, and ordinarily does not find novelty, and his originality consists only in finding the means to verify what the paradigm leads him to expect. We may ask, however, whether Kuhn's very definition of discovery as arising from the identification of a paradigm-violation does not restrict it too narrowly to one type of event within a broader range of possibilities. Elsewhere he tacitly acknowledges the existence of other types of discoveries, such as the "discovery of an additional element," which entailed no surprise, because they were predicted by the periodic table.[37]

Even if we place ourselves outside the orbit of Kuhn's definition of normal science, do Jacob's testimony and the cases that others and I have studied imply, nevertheless, that the route to a truly original discovery invariably includes one or more unpredicted events essential to the outcome? If research is unpredictable in a pervasive way, then we cannot even predict that the unpredictable will always take place. It must sometimes happen that an investigator guesses correctly at the outset the solution to the problem she is taking up, and that the initial means chosen to solve the problem succeed without requiring significant modification along the way. We might predict, however, that such cases will turn out to be exceptional rather than common, and all the rarer, the more novel the outcome is.

At the far reaches of scientific investigation, unforeseen observations or other types of events do occur that cannot be fitted into the matrix of previously acquired knowledge. In some Kuhnian scenarios they may cause the overthrow of the prevailing paradigm, or they may remain largely unfathomable and outside the scientific mainstream. None of the subjects of my studies operated in these realms. (Even Lavoisier, the leader of one of the acknowledged great scientific revolutions, considered to have overthrown the prevailing phlogiston theory, operated within broadly accepted rules for performing and interpreting chemical operations. Elsewhere I have argued that the contest between Lavoisier and Priestley that shaped much of the dynamic of the chemical revolution can be viewed as a competition between two contemporary research programs, rather than the overthrow of an established paradigm.)[38] The generalizations drawn from these studies do not, therefore, encompass the whole range of scientific investigative activity. They are, I believe, however, representative of a large proportion of the kind of activity that has for many generations of investigators driven science forward.

Part Four The Fine Structure of Experimental Investigation

Chapter 10 The Interplay between Thought and Operation

The eminent cognitive scientist Herbert Simon wrote in 1977 that "scientific discovery is a form of problem solving, and . . . the processes whereby science is carried on can be explained in the terms that have been used to explain problem solving."[1] Simon and others who have studied human problem solving experimentally have often been able to follow the mental processes through which persons given problems solve them during time intervals ranging from milliseconds to minutes. If scientists typically solved problems they encountered in the course of their investigations over comparably brief periods, the processes through which they have done so historically would, with few exceptions, be beyond the reach of historians. Only on the rare occasions when a scientist himself has recalled a memorable experience of this type, or when we come upon a document whose internal character suggests that the scientist may have written down a train of thought nearly as soon as it occurred to her, can we hope to recapture such intimate moments of creative activity. Because scientists, like most other people, are thinking continuously as they work, even such occasional captured moments can provide only extremely sparse samples of the thought processes that accompany an extended investigation.

It is not self-evident, however, that historians would benefit if it were possible to have a more complete record of what Gruber has called the "incredibly swift" stream of thought. The finest texture of such thought may not reveal the most meaningful contours of the intellectual activity of an investigator, any more than the observation of every grain of sand can convey the landscape of a dune. Introspection suggests that daily thought ordinarily circles repetitively back on itself, with slight variations on previous thoughts, and only widely spaced modifications representing some degree of novelty. Stable changes in ideas, according to Gruber, evolve at a moderate rate. "The movement of ideas is far slower than the swift but transitory currents of the continuous stream of thought which serves as the 'carrier wave' of creative work."[2]

In the case of experimental scientists, the rate-limiting factor in their mental progress may be set not by the speed with which such thoughts can course through their brains, but by the pace with which they can translate useful thoughts into laboratory operations. Moreover, the most frequent stimuli to further mental advance are likely to be responses to the outcomes, expected or unexpected, of such operations. The most meaningful units of creative action may, therefore, be measured less in cognitive moments than in the duration of individual experiments.

This being so, we would expect the character of the interplay between thought and operation to be strongly influenced by the length of time necessary for the investigator to prepare, perform, and evaluate each experiment. The scale on which these interactions take place varies greatly, depending on the type of experiments. Some require months for preparation or execution; others can be set up and completed very quickly. Of the examples I have followed, the lengthiest were some of the distillation analyses of plants conducted by Claude Bourdelin in the Academy of Sciences during the seventeenth century. Believing that they could avoid the alterations of the distilled material caused by excess heat if they substituted the length of time the heat were applied for its intensity, seventeenth-century chemists often prolonged such distillations for weeks or months. The shortest ones I have followed were those of Hans Krebs, who set up and performed a full set of manometric tissue slice experiments every morning and afternoon and completed the analysis of his data each evening.

The length of the time intervals between successive experiments can be expected to shape the nature of the interactions between outcomes and experimental reasoning. The shorter the time between experiments, the less we would expect the investigator's point of view to change between one experiment and the next, and the more likely we may expect it to be that each successive experiment is planned in response to the immediate result of the preceding one, rather than by developments external to the daily activity in the laboratory. When experiments require extended periods for preparation and execution, or when other tasks intervene between experiments in a particular line, we would expect successive experiments more often to reflect, in addition to the experience of the preceding ones, some evolution in the thought of the investigator during the interval, or an influence impinging on the course of the investigation from some contemporary event in the field.

Because laboratory notebooks typically capture very incompletely, if at all, the reasoning underlying the design of the experimental operations recorded, or of the mental response of investigators to the results, historians who attempt to reconstruct their reasoning from the recorded activity can be aided in their interpretations by such considerations. In my experience, I have regularly found that in a series of experiments carried out during relatively short time periods, the dominant reasoning behind changes in design from one to the next can be inferred from a close examination of the results of the former one. The alterations in the conditions established for otherwise similar experiments can frequently be accounted for by the deficiencies revealed in the preceding ones. When the interval is longer, however, the relation between experiments on the same or closely related problems is less likely to be fully explainable in such terms, and the historian is well advised to be alert for some development extrinsic to the experimental series itself to explain the changes in design or approach.

Typically Lavoisier pursued a line of experimentation aimed at a specific problem within his network of enterprises intensely for a relatively short period of time before turning to another problem, or interrupting the work for reasons extrinsic to the investigation. Sometimes he resolved the immediate question he had in mind, but often he left it unsolved, returning to it again after periods of time ranging from months to years. Although he often asserted that he had performed "many" of whatever type of experiments he presented one or two examples of in his publications, his laboratory records suggest instead that he most often made do with a relatively small number, of which, in some cases, no single experiment was without significant flaws. These characteristics of his work were probably not reflections of low standards, but of the limited amount of time that he had available for his laboratory work, and of the difficulty and expense necessary to construct the necessary apparatus under the technical conditions of the time, as well as the frequency with which components of an apparatus, pushed to the limits of the materials available, leaked or otherwise partially failed. The nature of the interactions between his thought and the operations he performed was strongly shaped by such circumstances. We can readily illustrate these aspects of Lavoisier's scientific style by summarizing the three series of experiments that he carried out on the respiration of animals, the first group taking place in 1776, the second in 1783, and the last in 1790.

When he performed his first respiration experiment on an animal, sometime between April and October 1776, Lavoisier had already completed the successive operations on the calcination and reduction of mercury that had allowed him to conclude that the atmosphere is composed of two portions, only one of which takes part in combustion processes. He had not yet named this component oxygen, but we may, for simplicity, anticipate the term he gave it three years later. He had by this time a well-developed experimental system for examining the effects of a combustion process on the air in which it took place. The apparatus consisted of a glass bell jar inverted over mercury contained usually in a marble basin. If, as he suspected, such a process produced fixed air, he could absorb the latter in a dish of caustic alkali floated on the mercury. The resulting decrease in volume became a measure of the oxygen consumed. The apparatus and procedures were adaptable with little modification to examine the effects of the respiration of a small animal placed under the bell jar, and it is not surprising, therefore, that in the first of this type of experiment that he performed, Lavoisier attained a significant result.[3]

The sparrow confined within the bell jar appeared at first only a little drowsy, but its respiration quickly became labored, and it died within less than an hour. The volume of the air diminished by only about one sixtieth, but when Lavoisier passed a portion of it through caustic alkali, the latter precipitated, and the volume decreased by about one sixth. He interpreted these results to mean that respiration both consumed oxygen and produced fixed air. Knowing immediately that this was a significant advance in understanding the nature of respiration, Lavoisier and his colleague Trudaine de Montigny carried out during October a series of similar experiments to confirm and extend the initial conclusion.

In the first of these experiments they used a robin in place of a sparrow, gave it a pedestal to stand on, and placed the bell jar over water rather than mercury. The latter change allowed them more readily to absorb the fixed air. The results were similar to the earlier experiment, but they applied more extensive tests to the irrespirable air remaining at the end. A more significant variation in the next experiment was that they filled the jar initially with pure oxygen rather than common air. The air remaining afterward precipitated limewater as before, but the residue, unlike that in the previous experiments, retained the properties of oxygen. By simplifying the conditions, therefore, this experiment strengthened Lavoisier's conclusion that respiration consumed oxygen and produced fixed air. Aside from a few auxiliary tests designed to examine the cause of the deaths of the birds, this completed the experimental series. Lavoisier had done enough to establish to his satisfaction the qualitative nature of the effects of respiration. Although he did obtain numerical values for the oxygen consumed and the fixed air produced, he was not concerned to reach more accurate or reproducible quantitative results. It took him another year to reach the interpretation that respiration consisted of the slow combustion of carbon, which combined with oxygen to produce fixed air, but this resolution depended on the general development of his theory of combustion, and did not require further respiration experiments.

When Lavoisier performed, with his collaborator Pierre Simon Laplace, the next series of respiration experiments, in February or March 1783, their goal was to determine the quantity of fixed air an animal produces in a given period of time (Figure 17). For this purpose they required a quite different experimental setup. To increase the quantities involved, they used a larger animal, a guinea pig, and lengthened the time that it could survive by using an open system in which they supplied it with fresh air at the rate of thirty pints an hour. They measured the fixed air produced by passing air from the respiration chamber through two successive bottles containing limewater, a procedure that Lavoisier had utilized extensively in his studies of other forms of combustion. The disadvantage of this system, in comparison to the earlier experiments, was that they could not measure the oxygen consumed. That did not present an immediate problem, however, because Lavoisier now regarded his qualitative interpretation of respiration as well established, and needed only one experimental indicator to measure its rate. By making certain assumptions, such as that the volume of the oxygen consumed equaled that of the fixed air produced, he was able to calculate the consumption of the former from the measured quantity of the latter.

We might view this experiment as a direct continuation of the series broken off seven years earlier, if we were to assume that when he had established the qualitative exchanges comprising respiration, it would be a natural next step to examine the process quantitatively. Lavoisier's motives for doing so were, however, not that simple. The reason that he wished in 1783 to establish the quantity of fixed air produced in a given time involves the whole development of his larger research program over the intervening years, as well as other contemporary developments. It was in the context of the calorimetric method that he and Laplace devised to measure the heat released in various processes—a method

First page of record of a respiration experiment carried out by Lavoisier and Laplace in 1783.

Suitte de l'expérience cy contre

il resulte de cette expérience qu'une volume d'air
vital d'un pied cube ou de 1728 pouces s'est reduit
par la respiration en 1674, pouces 56.

qu'il s'est formé air fixe – – – – – – – 228. pouces 78
qu'il a été employé environ 53 pouces 44 d'air vital a faire de
l'eau que le gu que la quantité d'air inflammable evacué du poulmon
a été de 4 grains 715 gr.
qu'il s'est formé eau 31, 435 grains
il a été employé a faire de l'air fixe —159, 6816 pouces il s'est formé -11-
En calculant la quantité de l'air nitreux d'après la formule p.43

$$x = \frac{(a + b - c)\, 68}{108}$$ je trouve pour la quantité réelle d'air nitreux

1er experience	97.6	
2me ex – – – –	97.0	
3me – – – – –	96.5	
4 – – – – –	96.0	
5 – – – – –	96.0	
6 – – – – –	96.0	
7 – – – – –	95.0	

d'ou je conclus que 100 parties
de l'air vital originairement employées
contenoient 99 parties d'air vital
réel.

après l'opération l'air nitreux étoit sans doute au même
degré de bonté, mais l'air vital ne contenoit plus que 88 à 89
parties d'air vital réel.

Il resulterais de la ou que la respiration a fourni de la
mophette ou que l'alkali caustique n'avoit pas absorbé tout
l'air vital.

*Il a été extrait du poulmon 31, 0492 de matières charbo-
nneuses.

that was in part a response to experiments on animal heat recently published by Adair Crawford—that they wished to compare the quantity of fixed air produced by a guinea pig with the heat produced during the same time, in order to compare both quantities with the corresponding quantities released in the combustion of charcoal. The experimental method Lavoisier and Laplace now applied to study the respiratory exchanges was, thus, not devised only in response to the limitations in the results of the older experiments, but reflected also a substantially changed outlook on the problem, brought about by the passage of time.

The result that Lavoisier calculated for the oxygen consumed in the new experiment turned out to be so low that he did not believe it. Consequently he and Laplace reverted to an experiment in a closed vessel, more like the ones performed in 1776, in which they could measure the decrease in oxygen directly. They also reverted to birds. To increase the time in which the birds could survive, they filled the vessel, over mercury, with pure oxygen. This experiment gave more promising results. They tried, therefore, a similar experiment with the guinea pig, filling a larger bell jar with three times the volume of oxygen to accommodate its larger respiratory rate. The result they now obtained appeared to give them considerable confidence, but they had not given up on the initial plan to carry out the experiment in an open, renewable system in which the animal could breathe much longer. Combining this approach with that of the preceding experiment, they now passed pure oxygen instead of ordinary air through the open system. In a second, similar experiment, they added a new procedure, passing a large quantity of air through the system after they had removed the animal, to ensure that all of the fixed air had been swept out into the limewater bottles.

In this sequence of experiments, done within two or three months, the purpose for which Lavoisier and Laplace carried them out did not change. We can, consequently, see the investigators over this short, intense period of activity designing each experiment almost purely in response to the limits perceived in the preceding one. Remarkably, they did not settle on any one of these methods as optimal and repeat it several times, but stopped after carrying out one of each variety. The reason they apparently felt they could do so was that three of the differently conducted experiments yielded results sufficiently close to one another so that they chose an average result from these three, discarding the other experiments presumably as flawed. For reasons that I have suggested above, this tendency to get the most he could out of a few, sometimes brilliantly conceived, but still problematic experiments, rather than to continue experimenting until he could be certain of obtaining unproblematic results with a standardized method, was typical of Lavoisier's experimental style.

In 1785 Lavoisier modified his theory of respiration to include not only the combustion of carbon, but also the formation of water from inflammable air and oxygen. Retrospectively he attributed the cause of this change to a deficit observed between the quantity of oxygen consumed in the respiration experiments of 1783 and the amount that could be accounted for in the fixed air formed. A close analysis of the several stages of the calculations he made at the time reveals, however, that he did not find this deficit, which was small enough to fall within the expected limits of error in such experiments, until he had already decided by analogy with other processes he had been studying, such as the combustion of charcoal and of wax candles, that respiration, too, produced both water and fixed air. Further typifying the manner in which he extracted the most he could from the results of experiments already performed, he then went back to one of the experiments of 1783 in which he and Laplace had placed the guinea pig in oxygen in a closed system, and reworked the data until they fit his newer conception of respiration.

When Lavoisier returned to respiration experiments in 1790, his approach to the problem had changed in many ways from what it had been in 1785. Now regarding the basic nature of the process as settled, he was interested mainly in the effects of varied conditions on the respiratory rate. Requiring again that the experimental animal be able to survive for a long period, he had to design an apparatus in which the oxygen consumed could be renewed, and the fixed air (now called carbonic acid) could be continuously removed. Because his newer theory of respiration assumed the production of both carbonic acid and water, he could no longer use the formation of the former as the common measure of the respiratory rate. Because oxygen took part in both processes, its consumption was the appropriate measure. To measure its consumption in an open system required new eudiometric methods, whose development he left to his young assistant in these experiments, Armand Seguin. From 1777 until at least 1783, Lavoisier had regarded respiration as a slow combustion whose primary function in animals was to supply animal heat. By 1790 he had come to view respiration also as the source of work, and accordingly designed experiments on human respiration (for which Seguin served as the subject), in which he measured the increase in the rate when a person lifted a weight a given vertical distance. Thus, although similar to his earlier respiration experiments in general principle, those he began in 1790 differed from the previous ones in almost every operational detail, as well as in overall purpose. None of these differences can be accounted for as responses to deficiencies or limits in the earlier experiments. Almost all of them can be ascribed to changes in Lavoisier's outlook on the question that took place during the seven years separating the two series of experiments.

It can often be especially instructive to follow the interactions between the reasoning and operations of an investigator during a period in which things are not going well. Lavoisier found himself in such a situation in 1787, when the entire thrust of his current research program pointed him toward the necessity of determining the composition of sugar, yet that substance resisted all his attempts to produce a satisfactory analysis. During the spring of that year he had arrived at a general conception of the composition of plant substances of great historic significance. "One ought to regard plant substances as a triple combination of oxygen, hydrogen, and carbon, a combination of little solidity that a very slight heat can alter and destroy. Then the oxygen and hydrogen combine with one another and form water, which is disengaged, carrying along with it a little hydrogen and carbon combined in an oily or soapy state, and the carbon stands alone." The model substance that underwent such changes he had in mind at the time he wrote down this speculative statement was sugar. He was, in fact, at the time oscillating between the view expressed here and an alternative conception of sugar as composed of carbon and pre-formed water. "The most rigorous analysis of sugar," he had found, reveals nothing but carbon and water or its elements.[4]

To go beyond such general statements required an analysis of sugar that revealed not only what its elements were, but in what proportions. That problem became increasingly urgent to him in 1787, because he was focusing his attention increasingly on fermentation, the operation by which sugar is converted to alcohol and carbonic acid. Of these three participant substances, Lavoisier had obtained good results for the composition of the two products, and needed only that for sugar to be able to account for the balance of the elements in the process. The problem was that, whereas the former two were easily combustible in air, sugar was very difficult to burn completely.

His understanding of combustion allowed Lavoisier to enumerate "four principal ways to oxygenate plant and animal substances. One can oxygenate them in free air; by distillation over an open fire, with the aid of the water they contain; by any fermentation . . . by their combination with the acids to which oxygen is only weakly bound, such as nitric acid or oxygenated muriatic acid."[5] The recalcitrance of sugar to the most direct method drove Lavoisier to try variations of each of the others on the substance.

In June Lavoisier wrote down in his notebook that he had attempted to combine sugar with concentrated sulfuric acid, from which he would expect to produce some hydrogen gas and sulfurous acid (the product of the reduction of the sulfuric acid). The carbon should be left in the bottom. When he carried out such an operation, he was able to collect some carbonaceous materials, but it was difficult to identify the sulfurous acid formed, and he collected only a small amount of impure hydrogen gas. The result seemed to produce mostly confusion. Turning next to distillation, he tried three times, adjusting the quantities of sugar and the size of the retort used each time to try to overcome difficulties encountered in the preceding attempt. He succeeded only in producing a syrupy acid of doubtful identity, some mixtures of gases, and some carbonized sugar. Somehow, however, by procedures of which no record survives, he was able from the combined results of these disparate experiments to derive an estimate of the proportions of carbon, hydrogen, and oxygen in sugar.

Meanwhile, Lavoisier had carried out, between April 27 and June 12, another of a series of fermentation experiments, using this time a simple stoppered flask with a tube that allowed the gases to escape. Assuming from the qualitative result of an earlier fermentation experiment, in which he had captured the gases, that the only such product was carbonic acid, he calculated the quantity produced in the new experiment by the difference between the weight of the flask and its contents at the beginning of the experiment and that at its end. Having determined the proportions of carbon and oxygen in carbonic acid several years earlier, he was now readily able to calculate the quantities of those elements produced. To determine the quantity of alcohol produced, he subjected the contents of the flask at the end of the experiment to a distillation, and a complex analysis of the several fractions received. Knowing also the composition of alcohol from earlier combustion analyses of this readily burnable substance, he now knew the total quantities of the elements contained in the products of the fermentation.

At this time Lavoisier believed that, in addition to the sugar, some water was decomposed in the fermentation process. From the overall change in the weight of the flask, he was able to determine the amount of water so consumed, and from his knowledge of its composition, the quantities of hydrogen and oxygen. Of the substances participating in the process, therefore, the only one whose composition he had not been able to establish firmly was the sugar. Now he was able to achieve indirectly what had eluded him directly, calculating the proportions of hydrogen, oxygen, and carbon in the sugar consumed from the quantities necessary to balance his equation for fermentation. This result did not diverge sharply from the "experimental" result that he had somehow cobbled together from his efforts to combine sugar with sulfuric acid and to analyze it through distillation, a convergence that temporarily afforded him considerable satisfaction. He composed, and probably read to the Academy in November 1787, a memoir on fermentation in which he announced these results as the climactic test of his whole new system of chemistry.

Underneath the air of confidence with which he conveyed this outcome, however, Lavoisier was aware of the fragility of his claims. He did not publish his memoir, and in the spring of 1788 he renewed his effort to find a better method to analyze sugar. Having already tried each of the four strategies he had earlier been able to imagine for the purpose, he now went beyond that list to try heating the sugar with red oxide of mercury. The critical experiments on the reduction of the mercury calx without charcoal that had led him and Priestley to the discovery of oxygen twelve years before had been possible, it was now easy to reason, because of the exceptional ease with which the oxide gave up its oxygen. It thus appeared that the oxide might appear a suitable agent to oxygenate sugar. The first such experiment he tried went too rapidly to control, but after making adjustments in the proportions of the oxide and the sugar, he performed the operation smoothly and collected the carbonic acid formed. The outcome must have been a disappointment to him, however, for the carbon content calculated from this result was less than a third the proportion he had calculated for sugar from the fermentation experiment.

This setback seems to have persuaded Lavoisier to postpone efforts to perform a complete analysis of sugar and concentrate instead on finding a way to determine the carbon content that would conform more closely to his expectations. Returning to his earlier method of combining the sugar with sulfuric acid, he was able to collect, calcine, and weigh the carbon produced in the bottom of the retort. It turned out to be 28.8 parts per hundred of the sugar. Assuming that if the calcinations had been complete, the carbon would have lost a little more weight, he corrected the result to exactly 28 parts per hundred, a figure he obviously favored.

We need not follow further in detail the tortured last stages in the pathway of Lavoisier's attempts to analyze sugar. Returning once more to red oxide of mercury, which he hoped might be developed into a general method for oxygenating plant and animal matters, he heated sugar again with it, collected the carbonic acid produced, weighed the oxide at the beginning and the mercury at the end to determine the quantity of oxygen that it had supplied to the sugar, and, during the complex calculations he imposed on his data, reached a result for the proportions of carbon and oxygen in carbonic acid that diverged drastically from the figure he had used for four years, based on his extensive earlier experiments to establish those proportions. When he used the new figure in the calculation of the elementary proportions of sugar, however, he reached an outcome close enough to the "calculated" composition from the fermentation experiment, so that for several weeks he abandoned his old values for carbonic acid and worked with the value computed from this single anomalous result. Eventually he recognized the imprudence of this decision and reverted to his older proportions (72 parts oxygen to 28 parts carbon, which happens to be very close to the modern value). In August 1788, Lavoisier searched once more for another oxygenating agent, trying black oxide of manganese, but without reaching a satisfactory result. When he published his famous balance sheet for fermentation in his *Traité de chemie* a few months later, he included proportions for the composition of sugar that he stated to be the outcome of a "long series of experiments carried out in different ways and repeated many times." Left unstated was that none of these experiments had been very successful, and that he must have arrived at the published proportions by making multiple assumptions and adjustments to reconcile the many inconsistencies he had encountered.

Although we must judge that Lavoisier failed in his quest for a reliable, accurate method of analyzing sugar, we can at the same time note the resourcefulness with which he went about the task, the extent to which he was guided by the deep understanding of the nature of combustion that his own prior work had established, and his flexibility in adapting to setbacks and trying new possibilities. In his search for an agent that would readily yield to a plant or animal matter the oxygen that these substances would not absorb directly from the air or from oxygen gas, Lavoisier had opened the investigative pathway that led ultimately, in the hands of his successors, to the success that had eluded him. Those achievements, produced by Joseph-Gay Lussac, Jöns Jacob Berzelius, and finally Justus Liebig in the early decades of the nineteenth century, drew on advances in the design of chemical apparatus, and techniques for constructing apparatus and joining their components in airtight linkages, as well as knowledge of the power of various oxidizing agents, that were not yet present in Lavoisier's time. In other analytical efforts in which he had been auspiciously successful, Lavoisier had often pressed the materials available for building apparatus to their limits. In his contest with the composition of sugar he pressed beyond those limits. The fine structure of the engagement of Lavoisier's far-reaching thoughts, confronted with obstacles he could not surmount under the conditions in which he was constrained to operate, reveals the creative capacities of a great scientist as fully as do the moments of his triumph.

In contrast to Lavoisier, who was able to perform relatively few experiments of a given type, and who extracted, therefore, as much information as he possibly could from the data yielded even by flawed operations, Hans Krebs performed multiple experiments day after day, year after year. He seldom hesitated to repeat and vary experiments whose outcome was uncertain or inconclusive, altering the conditions to improve their clarity or to bring out more fully effects that were unexpected. Because he also typically planned ahead only one day at a time, one can very often interpret the experiments of any given day in terms of his immediate response to the results of the work of the preceding day. Because he also kept in close touch with the current literature in his field and sometimes interrupted a line of investigation to follow up a lead suggested to him by a recent article, however, we also find not infrequently in his notebooks experiments that appear unrelated to what had come just before. In almost every case it is possible to identify the outside source of such a digression. These characteristics of his experimental style were partly an expression of his personal temperament: his disciplined desire to use all of his time effectively and his conviction that it was better to go ahead and test an idea experimentally rather than to ponder extensively whether it was worth doing. The effectiveness of this style depended also, however, on the special capacities of the experimental system that he had inherited from Warburg. It was the precision of the Warburg manometric tissue slice methods, the rapidity with which experiments could be set up and performed, and the ease with which one could recognize a significant or unexpected result, that provided the flexibility to proceed in this manner. We can illustrate these features of the fine structure of Krebs's investigative pathway at almost any point along its trajectory during the decade in which he himself performed experiments alone at the bench or with the assistance of one or two students. It will be convenient to do so by amplifying the summary description already given in chapter 8 of the first segment of the episode, beginning in January 1936, that led him from his prior concern with amino acid metabolism back to the problem of oxidative carbohydrate metabolism.

While preparing a review article on the metabolism of amino acids and related substances, Krebs reexamined a mechanism that Franz Knoop had proposed in 1910 for the synthesis of amino acids. One molecule of pyruvic acid would react with a second keto acid and ammonia to produce an acetyl amino acid intermediate plus CO_2. In a second step, the acetyl compound would yield the amino acid plus acetic acid:

$$R\text{-}CO\text{-}COOH + NH_3 + CH_2\text{-}CO\text{-}COOH \rightarrow R\text{-}CH\,(NH\text{-}CO\text{-}CH_2)\,COOH + CO_2 \rightarrow$$
$$R\text{-}CH\,(NH_2)\,COOH + CH_2\text{-}COOH$$

Believing that the fact that the formation of the amino acid did not require the presence of oxygen constituted "strong support for the scheme," Krebs decided to test it in his experimental system. This was typical of the way in which he initiated new episodes along his research trail. Not a deep theorist himself, he often found leads for experiments in reaction mechanisms he read of in the literature. Having previously ventured a few unsuccessful efforts to synthesize amino acids from keto acids, he thought that Knoop's scheme might offer a more promising route. The ease with which the manometric tissue slice method could be adapted to test any such mechanism encouraged him to try out the idea without delay.[6]

To begin, Krebs prepared the acetyl compounds of two amino acids, acetylglutamic acid and acetylalanine. Using the Van Slyke method for measuring amino-nitrogen, he could easily determine whether the addition of either of these compounds to the medium of a tissue slice would cause an increase in the amino-N present, which he would interpret as an indication that the second step in Knoop's reaction scheme was occurring in the tissue.

In the first experiment, in which he included only acetylalanine, with and without pyruvate, Krebs found only a "small effect," no larger than the effect of a control using pyruvate alone. For the next experiment he included both acetylalanine and acetylglutamic acid, each with and without pyruvate, and several controls, including pyruvate + ammonia. Because, according to the Knoop scheme, one of the two keto acid molecules involved in the first step was always pyruvic, but the other might be either a second molecule of pyruvic or a different acid, pyruvate alone could be expected to take the part of both of them. If the acetylamino compounds were intermediates, then they ought to form amino-nitrogen at a rate at least as large as the overall reaction. The outcome was equivocal. Pyruvate + ammonia produced a far larger effect than any of the runs including the acetyl intermediates. In the face of such unencouraging initial results, Krebs often quickly dropped such a new venture and returned to a main line of investigation. In this case, either because he was not at the time engaged in another main line, or because he regarded the Knoop proposal as too promising not to pursue, he continued.

On January 17 Krebs compared the formation of amino-nitrogen from acetylalanine with that from pyruvate + ammonia in two rat tissues. In liver tissue the "hydrolysis of acetylalanine [was] certain, but slower than amino formation with NH_4 + pyruvate," another ambiguous result. In kidney tissue, on the other hand, there was "enormous splitting" of acetylalanine. It appeared that the nascent investigation had yielded its first strong positive signal.

During the following ten days Krebs obtained similar evidence for the hydrolysis of acetylglutamic acid in kidney tissue. Testing other tissues gave mainly negative results, but Krebs was by this time convinced that the second step in Knoop's scheme did take place in the kidney, and turned his attention to the first step. On January 31, he carried out an experiment using only pyruvate and ammonia, measuring the production of CO_2 that would be expected to form in the first step, as well as amino-N, expected to form in the second step. Although the increase in both quantities was very modest, the experiment was mildly supportive of the scheme. Moving to extend the evidence to instances involving two different keto acids, Krebs next tested pyruvate + ammonia + ketoglutaric acid. In this case he obtained an increase of CO_2 production of 94 percent with pyruvate and ammonia, and of 311 percent with ketoglutarate and pyruvate. These results provided dramatic support for the Knoop scheme, and Krebs must have felt that he was on the way to the solution of a long-standing problem.

Turning next to a third potential reaction of the same type, acetoacetic + pyruvic acid + ammonia, Krebs observed an even more impressive result. Whereas pyruvate + ammonia increased the rate of CO_2 production 244 percent over the control, pyruvate + acetoacetate + ammonia raised it by 730 percent, an amount he put down as an "enormous increase." At first sight this result appeared to be a strong confirmation of yet another variant of the Knoop mechanism. At this point, however, Krebs asked himself the question, "Is NH_3 necessary?" The reason for his doubt was not only that he had omitted to run the control pyruvate + acetoacetate without ammonia, but because he had thought of an alternative reaction, not involving ammonia, that might also account for the large production of CO_2 that he had observed. He wrote it down:

$$CH_3 - CO - CH_2 - COOH + CH_3 - CO - COOH + H_2O \rightarrow$$
acetoacetic acid + pyruvic acid

$$\rightarrow CH_3 - CHOH - CH_2 - COOH + CH_3 - COOH + CO_2$$
β-hydroxybutyric acid + acetic acid

It is not evident whether he found this reaction in the literature or whether he simply devised it by means of "paper chemistry." It shared with the Knoop mechanism two of the same products, and was similarly a type of dismutation mechanism that required no molecular oxygen. Unlike the Knoop mechanism, however, this one had nothing to do with the synthesis of amino acids. To take it up, therefore, was to turn away from the problem with which he had begun, in order to explore another possible line of investigation. After making the control test that he had previously left out, and finding "no (definite) effect of ammonia on *this* reaction" (that is on the specific reaction of acetoacetate with pyruvate), Krebs took that turn. After one more experiment in which he added a method to measure the decrease in the amount of acetoacetate present and again found no effect of ammonia, he abandoned for now the study of amino acid metabolism to concentrate on the unforeseen possibility that had arisen in his path. The incentive for him to do so can be inferred from some of his earlier interests. He had several times in the past sought without success to illuminate the role both of pyruvic acid and acetoacetic acid in metabolism. Now he saw a favorable opportunity to link these two compounds together through their participation in an example of the general class of dismutation reactions that were of considerable current interest in the field. A further attraction was that one of the products, acetic acid, had also long been known to be central to metabolic pathways, but had not so far been fitted into place.

On February 14, Krebs posed a crucial test for his new hypothesis. Repeating the experiment, he applied a lanthanum test highly specific for acetic and propionic acid, to see whether acetic acid had formed in the tissue. The result was, he wrote down with emphatic notation, "negative !!!" The next day he tested for the formation of the third product, β-hydroxybutyric acid, and obtained a positive result. These conflicting outcomes probably appeared to Krebs a partial setback, indicating not that he was on a wrong track, but that the phenomena he was observing were probably more complex than could be encompassed in the single equation he had posed.

When confronted with such situations, Krebs typically loosened his experimental approach, exploring various combinations of the substances he had been working with in the hope of a clue that would indicate which direction to turn next. Testing such possibilities as the "influence of acetate on the disappearance of pyruvate," and the formation of ketone bodies from acetoacetate and pyruvate, he began to notice that the control experiments with pyruvate alone seemed regularly to give rise anaerobically to greater amounts of CO_2 than did its combinations with other keto acids. A background phenomenon gradually moved to the foreground, and he soon found himself studying the general problem of the "anaerobic disappearance of pyruvic acid in animal tissues."

Even though his most recent experiments appeared to be diffusing his initial focus on a single hypothetical reaction mechanism, leaving him searching for other leads, Krebs was confident that he was onto a fruitful course of investigation. In a progress report he supplied to his departmental chief, Edward Wayne, for a grant application about this time, Krebs wrote:

Preliminary experiments carried out in this Department with the tissue slice technique suggest that the primary reaction in the synthesis of amino acids from ketonic acids is the formation of an acetyl amino acid. This is then deacetylated by a specific enzyme. The energy necessary for the synthesis is provided by the oxidation of pyruvic to acetic acid and the actual transmission takes place by an intramolecular arrangement of dismutation type in an intermediate compound. Other biological reductions such as that of acetoacetic to β-hydroxybutyric acid can be explained on similar principles. Here an intermediate compound is formed with pyruvic acid and this breaks down with the liberation of acetic acid and carbon dioxide. This reaction is of special interest, representing as it does a link between fat and carbohydrate metabolism, and its study promises to shed light on the problem of ketogenesis and on the metabolism of the diabetic organism.

On March 7, Krebs digressed again from what had now become his main line of investigation, to take up the question of the formation of succinic acid in tissues. Although he had studied that question intensely during his first attempts in Freiburg and Cambridge to investigate the ways in which foodstuffs are decomposed in animal tissues, the occasion that brought him back to the problem was the latest publication on the subject by Albert Szent-Györgyi.

Elaborating on his earlier demonstration that fumaric acid can act catalytically to maintain the respiration of minced pigeon breast muscle tissue, Szent-Györgyi now proposed a scheme in which succinate and fumarate functioned as part of a hydrogen transport system through an exchange with a higher oxidation state. Succinate was oxidized to fumarate, but, he had found, fumarate cannot be reduced back to succinate in the tissues. Accordingly, he proposed that fumarate is instead further oxidized to oxaloacetate, and succinate is formed directly from the latter by a process of "overreduction."

The work and views of Szent-Györgyi strongly impressed Krebs. He thought that the new scheme was significant, but probably not quite right, because he could not make sense of Szent-Györgyi's conception of "overreduction" and thought that there must be some other mechanism through which fumarate can be converted to succinate. To begin his exploration of the situation, Krebs tried out a new method for the determination of succinic acid that had been devised by one of Szent-Györgyi's co-workers. The method worked well for him. He then tried, with only partial success, to confirm in his own experimental system several of the observations Szent-Györgyi had reported. Breaking off this effort, he applied the succinic acid method instead to answer a new question, whether "the conversion of pyruvic acid into succinic acid" can take place "in the absence of molecular oxygen." Thus what had begun as another diversion turned quickly instead into a new link between present and past investigative pathways.

At this point Krebs broke off his research to begin a journey to Palestine, prompted by efforts to recruit him for a position as director of a proposed division of a cancer institute in Jerusalem. When he returned a month later, he continued, in a similar manner, to explore these and further possible dismutation reactions explainable "on similar principles." As in the sample of his pathway described here, he oscillated between periods of sharp focus on specific reaction possibilities when the evidence seemed to point that way, and more diffuse explorations of broader ranges of possibilities when the evidence became more problematic. As described with less detail in chapter 8, he continued along the general lines laid out here through the summer and fall, and well into the spring of 1937, until his encounter with the paper by Martius and Knoop stimulated the decisive shift in direction from which the citric acid cycle quickly brought closure to his long endeavor.

One can see in this short episode many of the characteristics of Krebs's scientific style: his exploratory mode of operation, the ease with which his methods provided clues that either strengthened, weakened, or modified an initial idea, his readiness to digress from one problem to another without losing sight of his broader objectives, his ability to connect new possibilities to older problems, his openness to leads coming from the outside as well as from the results of his own daily laboratory activities, his well-attuned sense for the inevitable unpredictability of scientific research. His style was neither that of a visionary marching steadily toward distant goals, nor of one who thought long and deeply about what problems to pursue. Rather he was a disciplined improviser, responding quickly to ideas that he came across often by chance, as well as to unexpected observations, flexible enough to change directions opportunely, consistent enough not to become lost in the welter of possibilities that he encountered. As the segment we have followed in detail suggests, Krebs did not always appear inspired. When nothing remarkable turned up, he could appear to drift. He sometimes leaped quickly to conclusions he could not afterward sustain. His creativity was expressed more in his reactions to the unforeseen events that he encountered along his way than in the foresight with which he took each new investigative line. The magnitude of the successes that he attained in the long run, however, may lead us to conclude that such aspects of his personal style were exceedingly well adapted to the requirements for effective research in his field. The fine structure of Krebs's investigative pathway appears often scattered and sometimes hastily planned; yet it grew over the years and decades into a lifelong trajectory sustained with remarkable consistency.

The pace with which Meselson and Stahl were able to perform experiments in the investigation that led to their classic DNA replication experiment was limited by the time required for a run on the Model E analytical ultracentrifuge. In the early stages of their project only one such machine was installed on the Caltech campus. Demand for time on it was heavy and, in particular, because the equilibrium runs that their cesium chloride method required were unusually long, Meselson and Stahl sometimes had to wait in line for days or weeks for their turn. Consequently they often performed their early experiments in bursts of as many runs as they could fit in during an allotted time, working through the nights and weekends, separated by intervals during which they waited for another turn. The pattern of their experiments reflects the influence of this temporal rhythm. In a series of experiments carried out in close succession, one can see each one as a response to the preceding one, whereas after an interval they sometimes returned to pursue a somewhat different problem or sub-problem that had arisen in the meantime. To illustrate this pattern, we will follow a segment that started at the beginning of the 1956 Christmas holidays and continued with particular intensity through the first half of January 1957.

As described in chapter 9, the surprising discovery that cesium chloride formed a density gradient made it at first problematic for Meselson and Stahl whether the method they had envisioned to separate substituted from unsubstituted DNA would work. During the first few centrifuge runs they carried out in late October, they attempted to form bands with ordinary whole phage. At first they obtained nothing at all, but on the fourth equilibrium run, begun on November 4, the films showed a broad, diffuse band on the less dense end of the cell, and a narrow, somewhat sharper band near the denser end. They guessed that the phage was partially disintegrating, the diffuse lighter band representing phage, the denser sharp band representing DNA that had been released by phage that fell apart. To check this assumption they began, on November 24, a run in which they placed a known sample of calf thymus DNA. The latter "banded at almost the same place" that the narrower band had appeared in the preceding experiment, although this was only a rough guess, because they did not yet have a fixed reference point to compare the densities of different CsCl solutions. During the first week in December, Meselson reinforced the identification of these bands as DNA by attaining a similar result using DNA released from phage T4 through "osmotic shock." These preliminary results made them hopeful that DNA would form bands well enough defined so that it might be possible to distinguish those of different densities.[7] When they began the intense series of experiments over the Christmas holidays, their principal aim was to improve the resolution and reliability of their procedures sufficiently to realize that goal.

Using a specially purified suspension of T4 DNA, Meselson began, late in the evening of December 19, a run in which the DNA was placed in a CsCl solution whose density (= 1.69 at 23°C) was selected to match the calculated density of the DNA. After thirty-one hours the light region of absorption on the successive ultraviolet films had coalesced into a well-defined, though still rather broad band. A densitometer tracing showed a shoulder on one side of the band. Not yet certain how sharply their method could be expected to band homogeneous DNA, they may not yet have been able to decide whether this one contained some heterogeneity.[8]

On the day after Christmas, Meselson ran whole phage in a solution of density = 1.48, and obtained, as six weeks before, a broad band on the less dense portion of the cell and a sharply defined one at higher density. They probably interpreted the two bands, as they had the previous one like it, to represent whole phage and DNA released from the phage that had disintegrated. Two hours after this run ended, Meselson began another with released DNA alone, reducing the concentration to one quarter of that used previously. This time he obtained a nearly symmetrical band with a sharply defined peak, an encouraging sign that they were approaching the degree of resolution necessary. Because, however, the optical density of the solution was greater below than above the band, they suspected that the run had not reached equilibrium, and Meselson reran the same cell on December 29 at a higher RPM. This time they obtained a band sharper and narrower than any seen before. The next day Meselson performed a similar run with calf thymus DNA and obtained a band whose densitometer curve was almost exactly superimposable on that from T4.

Confident now that the resolving power of the method was adequate to the task, they turned to their next major goal, the separation of 5-bromouracil-substituted DNA from normal DNA. Stahl had already prepared stocks of T4 produced by infecting bacteria grown in a medium containing the substituted base, but the extent to which the 5BU had been incorporated into the phage DNA remained uncertain. The fact that only a few mutants appeared suggested that perhaps there had not been much substitution, but they optimistically went ahead, placing a mixture of "5BU" and "parental" DNA in the same centrifuge cell, in a CsCl solution of density = 1.71.

The run, which began on the afternoon of January 1, produced only a single, sharp band. What was it? On the centrifuge log Meselson wrote only, "mystery band." Displaced two-thirds of the way toward the denser end of the cell, the location might have fit the expected density (= 1.8) of 5BU-substituted DNA, but in that case where was the band representing the unsubstituted DNA? The alternative interpretation was that very little 5BU had been incorporated, and that the "substituted" and normal DNA had banded in the same place.

Thinking that they might be able to detect the substitution of 5BU into whole phage more easily than into the isolated DNA, they next ran together substituted and unsubstituted T4, but found no bands. With parental stock alone, Meselson then produced a single, sharp band. Becoming by now increasingly aware of the difficulty of telling "where" the bands were, when small variations in the density of the solution could readily cause large displacements, he attempted to provide a reference point by adding to the cell carbon tetrachloride, whose density was known to be 1.59. This and several similar attempts failed, however, to establish a useful density marker.

Turning to a stock of 5BU phage prepared differently, Meselson performed another experiment using substituted and unsubstituted DNA together, but again found only a single band, indistinguishable from the one produced a week earlier with ordinary T4 DNA. A similar experiment on January 9, using yet another preparation of 5BU stock together with normal T4 DNA, turned out no differently. The urgency of determining whether or not they had achieved any incorporation of 5BU into these stocks led Meselson now to run the new stock alone, without normal DNA. Using the same CsCl solution as before, he found this time that no bands at all formed. It was not the case, therefore, that the 5BU had banded at the same place as the unsubstituted DNA, but that something was preventing its bands from appearing. The first idea to occur to them was that perhaps the density of the substituted DNA was just enough larger than that of the most dense portion of the solution, so that the DNA was falling to the bottom of the cell.

To test this possibility, Meselson placed a concentrated solution of the 5BU DNA in a more concentrated CsCl solution (density = 1.779 instead of the usual 1.71). Late on the evening of January 13, he was able to see on the developed films a band that, although faint and diffuse, was located near the heavy end of the cell, about where one would expect 5BU-substituted DNA to form.

Now strongly encouraged that they had overcome the obstacles they had encountered, Meselson and Stahl moved quickly to their immediate goal, the separation of substituted from normal phage DNA. Simply adding ordinary T4 DNA to the same cell, Meselson reran it. On the new films the broad 5BU band appeared where it had before, and a sharper band showed up where normal DNA would be expected. By this time the two partners believed they were on the road to success. To James Watson, Meselson wrote shortly afterward, "We have banded 5BU DNA from a T4 stock which was about 60 per cent substituted as determined chemically. To the same centrifuge cell contents we then added some normal T4 DNA and found a very healthy separation of the two sorts of DNA. Plenty of room for finding intermediate pieces from transfer experiments. Also this mixed run looks just like the sum of separate runs each with only one sort of DNA. This relieves us of interaction worries." As the summary account in chapter 9 suggests, the road ahead toward a successful transfer experiment still held more twists and turns than Meselson and Stahl knew at this point. They had, however, during four weeks of nearly nonstop experimentation turned a vision into a working experimental system. Responding resourcefully to both the shortcomings and the leads provided by each experiment they performed, they had built the solid foundation that made everything that came afterward, the anticipated as well as the unanticipated, possible.

When carried down to fine levels of detail, these accounts of segments of investigative pathways become the stories of the daily lives of individual human beings. They show how such individuals manage their ordinary work, how they think about the events of the day, how they plan their next steps in response to what has just happened, whether foreseen or unexpected. They reflect the cognitive and emotional styles of particular persons, the opportunities and limits to their actions set by their particular professional positions, the resources available to them, their prior training, their access to centers of authority in a field, the degree of authority they themselves have attained. Each story is unique, local, and personal. To that extent I agree with the constructionist viewpoint that all scientific knowledge originates in local, contingent circumstances.

Told at a local level alone, however, such stories have little meaning. Each of these stories is situated within successively larger stories. Each episode is connected to everything that the investigator had previously experienced, including his apprenticeship within the field he entered. Each investigator had entered a field within a historical time and place that was in one sense local, but located at a particular point linked indissolubly with the broader scientific landscape, and integrated into a much longer temporal sweep; and that cannot be interchanged with any other historical context. None operated, even for a day or a moment, in isolation from the contemporary state of the field in which he practiced. None of the moves that we can follow him making from experiment to experiment could have led to recognized achievements had his thoughts and actions been based purely on local considerations. Each story is, therefore, at once personal and unique, and representative of features common to scientific research in general. Each story is a small piece both in the individual quest for a meaningful, well-spent life, and in the collective advance of science.

Chapter 11 Flashes of Insight
and Moments of Discovery

The "Eureka moment" epitomizes a common image of science: that its advances often come in moments of insight during which the creative scientist suddenly sees the solution to a problem with which he has been struggling; or, a new conceptual framework emerges all at once in his mind. The prototype story is based on the exclamation that Archimedes is supposed to have uttered in the bathtub when he saw how to determine the proportion of base metal in the crown of Hero from its specific gravity.

The details of Archimedes' Eureka experience are lost in the mists of early legend, but scientists of the more recent past have testified to similar experiences. The most famous of these are the account by August Kekulé of the dreamlike state in which he first envisioned atoms linked together in chains while riding on an omnibus;[1] and that by Henri Poincaré of his recognition, just at the moment when he put his foot on the step of an autobus, that the transformations he had used to define what he called "Fuchsian functions were identical with those of non-Euclidian geometry."[2] Charles Darwin testified that when he happened to read Malthus on *Population,* "it at once struck me that under these circumstances favourable variations would tend to be preserved and unfavourable ones destroyed. The result would be the formation of new species. Here, then, I had at last got a theory by which to work." Concerning the "tendency in organic beings . . . to diverge in character as they become modified," Darwin wrote, "I can remember the very spot in the road, whilst in my carriage, when to my joy the solution occurred to me."[3] More recently, the immunologist Niels Jerne has related that the idea for the selection theory of antibody formation came to him while he was walking home in Copenhagen, and James Watson has described the moment when, while tinkering with a mechanical set of the four bases contained in DNA, he became aware that they fitted together in two pairs, each of the same overall dimensions.

Such personal testimonies have often been taken as exemplary of the high moments in the creative process in science. Analyzing the classic examples of Archimedes, Poincaré, and several other less well-known cases, René Taton generalized in 1962 that "the study of various types of invention and discovery has shown us that after a long effort of reflection and research a discovery will suddenly flash into the mind of the research worker by means of a sudden illumination, the so-called *Geistesblitz*."[4] Thomas Kuhn was undoubtedly influenced by such images when he likened the origin of a new paradigm to a Gestalt shift. In *The Structure of Scientific Revolutions* he wrote, "the new paradigm, or a sufficient hint to permit later articulation, emerges all at once, sometimes in the middle of the night, in the mind of a man deeply immersed in crisis. What the nature of that final stage is—how an individual invents (or finds he has invented) a new way of giving order to data now all assembled—must here remain inscrutable and may be permanently so."[5]

If we take seriously that such moments may lie at the heart of scientific change, then a number of questions arise. Do such events happen frequently or only under very unusual circumstances? Do they happen only to those powerful intellects we sometimes call geniuses, or to creative people in general? Are such events truly instantaneous, or do they emerge over measurable intervals of time? Are they "holistic" events, in which a new picture emerges fully drawn, or is there a structure that spreads rapidly from some germinal hint to the full insight? Are the retrospective memories of those who have testified to the occurrence of such an event accurate accounts of what actually happened? In the many more cases for which we have no such personal testimony, does the nature of these moments of discovery elude historical recapture?

Following Poincaré, Taton believed that scientists could be broadly divided into two types of personality depending on the "emphasis he places on intuition.

While the intuitive relies mainly on his 'illuminations,' his flashes of genius to show him the most fruitful path, the logician prefers to follow a more rigorous method and a more austere and systematic road." Generalizing from Poincaré's belief that flashes of insight arise from work of the subconscious mind, and supported also by the view of the physicist Louis de Broglie that some kind of "crystallization" occurs "quite suddenly," so that the worker "perceives instantly and very clearly, and from then on perfectly consciously, the main outlines of the new concepts that were latent in him," Taton wrote that "this flash of thought . . . does not generally appear during periods of assiduous work, but rather during those of rest or relaxation. . . . Maturing slowly as a result of previous effort and the work of the subconscious, a discovery will suddenly appear at such times as the investigator's mind did not seem to be dealing with it." Although he qualified Poincaré's interpretation of creative work by remarking that it could be applied "in full" only to those areas of mathematics and mathematical physics in which "observations and experiments play no more than a very indirect role," Taton wrote that the stress on sudden inspiration and the role of unconscious work extended to "other aspects of scientific discovery." Taton's description, so heavily influenced by Poincaré's account, which he thought the "most penetrating attempt by any discoverer to explain the genesis of his discoveries," was an expression of a widely shared view. The great "moments of discovery" in science have appeared to be intimately associated with powerful, instantaneous experiences of illumination that occur when the reorganization of ideas carried out by the unconscious suddenly wells up into the conscious awareness of exceptionally creative thinkers.[6]

Gruber, who has closely reexamined the nature of Eureka experiences—or, as he sometimes calls them, "'Aha' experiences"—has argued that such events are more frequent than is often imagined, that they are not instantaneous, but have a structure that develops over a measurable interval of time, that they are less likely to represent complete ruptures with the past than is usually imagined, and that they are linked to events that precede and follow them like the crest of an ongoing wave. From his experience in experimental psychology, he infers that it requires at least several seconds to conjure a familiar image or to recognize a fragmented one, and that to become aware of a train of related ideas of any complexity at all would probably take longer still. The time required is sufficient so that an individual beginning to have an insight is likely to be aware of where it may be headed before it fully emerges, and able consciously to "steer" it in a preferred direction, or avoid a direction that feels unsafe.[7]

From his study of the species notebooks of Charles Darwin, Gruber con-

cluded that Darwin may have had something that felt like a novel insight as often as once a day. Many of these would have been ephemeral or relatively in-
significant, but Darwin's example suggests that Eureka experiences of some sort
are common occurrences among people who constantly think creatively. Moreover, by the time he read Malthus, Darwin had already several times written
down ideas that appear retrospectively as weak versions of natural selection. That
he remembered only the more forceful Malthus incident is, Gruber believes,
suggestive of the fact that memory ordinarily simplifies the past by conflating
within the single, most powerful of a sequence of similar experiences, events that
in fact had occurred repeatedly.[8]

Conceptual transformations that appear from a distance to have required an
abrupt rupture with a past way of thinking may, when we have a dense enough
surviving record, dissolve into a progression of smaller steps. Kuhn's view that
the origin of a new paradigm requires a sudden Gestalt shift appears to have been
conditioned by his belief that successive paradigms are necessarily in conflict at
some essential point, leaving no logical transition from one to the other. The
only way across the mental gap is through a sudden "flash" in which the incipient core, at least, of the new appears all at once. Among the prime examples of
such paradigm shifts that Kuhn invoked was Lavoisier's move from the phlogiston theory of combustion to his oxygen theory. Although he did not identify
a specific point at which Lavoisier may have experienced a Gestalt shift, Kuhn
wrote that, "after discovering oxygen, Lavoisier worked in a different world."[9]

Larry Laudan has criticized Kuhn's "holist picture of scientific change" on the
grounds that Kuhn was mistaken to view paradigms as structures so integral and
self-contained that one can be exchanged for another only in a single, global conversion process. Laudan views paradigms instead as reticulated structures, made
up of empirical information, theories, methodological rules, and values that are
separable. These components are "individually replaceable in a piecemeal fashion."[10] Laudan was concerned primarily not with the process by which the germ
of a new paradigm originates in the personal experience of an individual, but
with that by which the scientific community reaches consensus on whether to
adopt a new paradigm. The argument can, however, be applied also to the question of origins.

There is, in fact, among the many documents tracing Lavoisier's thoughts between the time he took up the question of airs in 1772 and the time he publicly
challenged the phlogiston theory of Stahl and of Priestley in 1777, no evidence
of a particular moment in which he made the transition from the old to a new
paradigm. Rather, for several years he found himself in a position containing

some elements of what would eventually become his new paradigm, but retaining also elements of the paradigm he would ultimately abandon. Elsewhere I have described this transition as a prolonged "passage," during which Lavoisier lived partially in two worlds. I have suggested that his mental "picture" was not entirely coherent during this period, because he was unable to join together the partial solutions he had achieved for some of the sub-problems he confronted into a unified, integrated theoretical structure.[11] Although this case seems in part to favor Laudan's view of reticulated structures that are replaceable piecemeal, one can also argue that during the interval in which Lavoisier was somewhere between the two paradigms, his position was unavoidable but unsatisfactory, and that once he had completed his passage, what he had was no longer a reticulated structure composed of separable pieces, but something closer to Kuhn's conception of an integral paradigm incommensurable with the one Lavoisier now opposed.

Gruber's analysis of Eureka experiences would lead us to expect Lavoisier to have undergone not the single momentous Gestalt shift that Kuhn postulated, but a series of smaller illuminations associated with the origins of various pieces of the theoretical structure he gradually built up. In contrast to Priestley, however, whose seemingly transparent narratives of his experimental adventures include numerous descriptions of unexpected observations that occasioned new insights, Lavoisier seldom revealed, in his published memoirs, the actual course of development of his thought. Typically he rearranged the historical order of his investigations to make the progression more logical, concealing in the process whatever moments of sudden insight he might have experienced. Nor did he reveal such experiences in more informal retrospective contexts. Does that mean that whatever immediate moments of insight and discovery he experienced are forever lost to us?

Most of them undoubtedly are. There are, however, within the rich collection of Lavoisier's unpublished documents, numerous traces of thoughts that have the appearance of being put down nearly as they occurred to him. Some of these involve clarifications and further developments of ideas associated with writing and revising the manuscripts for his publications. Others are informal memoranda tracing a stream of reflections that sometimes led Lavoisier to ideas he recognized as novel. Few of these capture the seminal stages of what later became critical pieces of his theoretical structure. In one case, however, I was able to locate a note that, from its internal character, appears to stand very close to the initial insight from which his theory of respiration afterward grew. A portion of this note has already been quoted in chapter 8. To show the primordial character of the ideas included in this passage, I will here quote the text in full. On the front of a folded piece of paper Lavoisier wrote:

Ideas

One has thought up until now that animals and plants absorb air through the lungs and through the trachea only to cause it to circulate in the animal or plant economy. It is certain that it is fixed there.

When an animal [is kept?] under a pneumatic apparatus there is absorption of air. One sees it by means of the barometer placed under the receiver.

Some scientists have said that the air merely loses its elasticity, that is to say that it ceases to be in the vapor state and that it enters a combination.

Couldn't one surmise ~~that there is only a certain portion~~ [Lavoisier crossed out these words, replacing them with] that the heat of animals is sustained by nothing else than the matter of fire which is disengaged by the fixation of the air in the lungs.[?] It would be necessary to prove that whenever there is absorption of air there is heat.

On the inside fold he added:

But isn't the air itself composed of two substances, of which the lungs bring about the separation ~~and only absorb~~ [he crossed out these words and finished] of one of the two.[12]

The format, as well as the content, of the note is suggestive of the newness of this idea to Lavoisier. The bare heading "ideas" seems to imply that as he began, he was not even sure what topics his note would cover. The tentative nature and simplicity of his own "surmises," put in the form of queries rather than statements, suggest that he had not yet pondered them extensively enough to evaluate their significance or likelihood. The emendations he made as he went along suggest that he may have been formulating these ideas even as he wrote.

We cannot infer from this note the immediate setting in which Lavoisier wrote it. Even the date is unknown. Was he at the time at work in his laboratory, or in his study? Had he returned from a walk with his head full of ideas he hastened to put down, or had he awakened from a reverie and sought to make clear what had flitted, like Kekulé's atoms, through a relaxed mind? We have no clue. We cannot recover the affective state or the coincidental circumstances in which these thoughts occurred to their author. In compensation, we appear to have something more immediately linked to the primitive intellectual nature of the insight than we do in the well-known testimonies of Kekulé or Poincaré, which were filtered and mediated through their vivid, but necessarily temporally more remote memories.

The notebooks of Claude Bernard contain numerous entries that have the appearance of insights that he wrote down as they occurred to him (Figure 18). Here is a representative example. During November 1846, Bernard performed an experiment entitled "Influence of the Vagus Nerves on Nutrition." On a rabbit he sectioned these nerves and immediately afterward injected a glucose solution under the skin. Before the operation the rabbit's urine had been turbid and alkaline. Afterward it became transparent and only slightly alkaline, and contained large quantities of sugar. Having previously observed that the urine of fasting rabbits is clear and acidic, whereas that of the animals when digesting herbivorous food is turbid and alkaline, Bernard inferred that the cessation of the function of the vagus nerves had prevented the nutritive assimilation of the sugar injected and rendered the animal in a condition akin to fasting. This result inspired in him a train of ideas that ranged far beyond its immediate interpretation. Following his description of the experiment, he wrote:

> The sugar was no longer burned in the lung after the section of the vagi, because one has recovered it in the urine after the removal of these nerves—and it is because of this lack of combustion that it had not rendered the urine alkaline.
>
> Reflection. The lung is thus the organ in which the chyle burns; in which in a sense it is digested. One might say that in the digestive system it is the stomach that destroys the substances—and in the assimilative system it is the lung that destroys the chylous substances and assimilates them. Now it is a curious fact that it may be the same nerve, the vagus, which presides over these phenomena in the lung and in the stomach. And we have seen that this nerve abolishes in these 2 organs the phenomena that one could call more especially vital, whereas those that are purely chemical, such as the lactic transformation in the stomach, the absorption of oxygen in the lung continue without the nervous influence, whereas the vital phenomena cannot go on. When one sees the relation that joins the stomach and the lung which concur to the same end (nutrition) in this sense, that the lung operates again on that which the stomach has already caused to undergo a first elaboration; one is no longer surprised, I would say, at the nervous union that exists between the stomach, the lung and the heart.[13]

This passage obviously includes a more complex set of ideas than the comparable passage of Lavoisier's discussed above. In the course of writing it down Bernard gave considerable development to whatever may have been his primordial insight. Not all the elements of the "reflection" were new. That sectioning the vagus nerves caused digestion to cease he had already concluded from earlier experiments. What this experiment added was evidence that the same operation also prevented the assimilation of glucose injected directly into the circulation. That it might do so must have occurred to him prior to performing the experiment, because its design implies that he intended it to test

Page from Claude Bernard's laboratory notebook describing the beginning of a series of experiments on gastric digestion.

such an idea. What then seemed to have struck Bernard as a novel insight, worth putting down as a reflection? Most likely it was the sense that these two effects of sectioning the vagus nerves fitted together into a broad new picture of the nutritional process. That was an integration of earlier thoughts that came to him as a result of the experiment. The length of time that passed between the initial intimations of this picture and the written passage cannot be closely established, but the possibilities range from a few minutes to several hours. We cannot, therefore, identify Bernard's written reflection with an immediate Eureka experience, but we can infer that the passage came soon enough afterward to bear discernible traces of such an experience.

Bernard's insight did not survive the outcome of several more experiments that he carried out in order to test it further. That is true of almost all of the nascent ideas that one can find in his laboratory notebooks, and fits Gruber's view that creative scientists experience frequent flashes of insight, the great majority of which do not turn out to be the origin of a major scientific advance.

Among those flashes that did initiate historic advances, the account related by James Watson is particularly persuasive. Having decided to construct helical models with the phosphate backbones on the outside, he and Francis Crick were concerned with how the bases could be fit together on the inside in a manner that would make the outside backbones completely regular. For some time Watson was preoccupied with the idea that two like bases would be held together by hydrogen bonds. He came up with a scheme that looked good to him until Jerry Donohue pointed out to him that the hydrogen bonds he had chosen represented the wrong tautomeric forms of the bases. To assist him in trying other possibilities, Watson cut out of stiff cardboard representations of the physical dimensions of the bases. Arriving the next morning at the office he and Crick shared, Watson

> quickly cleared away the papers from my desk top so that I would have a large, flat surface on which to form pairs of bases held together by hydrogen bonds. Though I initially went back to my like-with-like prejudices, I saw all too well that they led nowhere. When Jerry came in I looked up, saw that it was not Francis, and began shifting the bases in and out of various possibilities. Suddenly I became aware that an adenine-thymine pair held together by two hydrogen bonds was identical in shape to a guanine-cytosine pair held together by at least two hydrogen bonds. All the hydrogen bonds seemed to form naturally; no fudging was required to make the two types of base pairs identical in shape. Quickly I called Jerry over to ask him whether this time he had any objection to my new base pairs.
>
> When he said no, my morale skyrocketed.[14]

That this was an illumination that Watson was likely to have *seen* all at once is made more compelling by the fact that the pattern he recognized was visually present before him. Formulating the insight in words may have been an afterthought. Of course, there may have been a brief temporal structure even in the emergence of the visual awareness. We are not told whether Watson just happened to put the bases together in this way and recognized the pattern only after he saw it, or whether some prior intimation induced him to put them together so as to confirm what he had just thought. In any case the process need not have occupied more than a few seconds.

This "moment of discovery" was, however, not detached from what came before, but fit as a nodal point along a continuing process of thinking about the ways in which the bases might fit together. It was the crest of a wave. Unlike the famous prototypical Eureka experiences, this one happened to Watson not while he was away from his work, but in the midst of concentrated study of the very problem his insight solved. Perhaps this case should modify the stereotype that Eureka experiences are most likely to happen during interludes when the thinker has put aside his work and allowed his problem to incubate in the unconscious regions of his mind.

The "idea of a selective mechanism of antibody formation" did occur to Niels Jerne, according to his own later testimony, while away from work, "one evening in March, 1954, as I was walking home in Copenhagen from the Danish State Serum Institute to Amaliegade."

> The train of thought went like this: the only property that all antigens share is that they can attach to the combining site of an appropriate antibody molecule; this attachment must, therefore, be a crucial step in the sequences of events by which the introduction of an antigen into an animal leads to antibody formation; a million structurally different antibody-combining sites would suffice to explain serological specificity; if all 10^{17} gamma-globulin molecules per ml of blood are antibodies, they must include a vast number of different combining sites, because otherwise normal serum would show a high titer against all usual antigens; three mechanisms must be assumed: (1) a random mechanism for ensuring the limited synthesis of antibody molecules possessing all possible combining sites, in the absence of antigen, (2) a purging mechanism for repressing the synthesis of such antibody molecules that happen to fit to auto-antigens, and (3) a selective mechanism for promoting the synthesis of those antibody molecules that make the best fit to any antigen entering the animal. The framework of the theory was complete before I had crossed Knippelbridge.[15]

Jerne's biographer Thomas Söderqvist ranks this account among the "classical *eureka* stories in the history of science literature." He also points out that, written ten years afterward, this was not a direct memory of the event, but Jerne's attempt to "make a logical reconstruction of the reasoning leading to the theory."

His recapitulation of his train of thought is, in fact, more complex than the first handwritten statement of the theory that Jerne composed in August of that year. It must, therefore, represent not a momentary flash of insight, but rather a sequence of conscious reasoning that followed some primordial intimation that the reconstruction does not necessarily capture. Assuming that Jerne's reconstruction does resemble in general, if not in its specific logical order, the train of thought that had occurred to him on that evening, how long might it have taken for him to formulate mentally a theory comprised of at least eight component ideas articulated with one another? In this case we can set a maximum duration by the time taken to walk from the Serum Institute to the Knippelbridge, and it might be rewarding to retrace Jerne's footsteps to obtain some sense of that magnitude.

Söderqvist found that at least one important detail of the account, that the event took place in March, could not be correct, because one of the preconditions was Jerne's discovery of an antibody in natural blood, an event that did not take place until later that spring. Agreeing with this point, Jerne later told his interlocutor that he now believed that the walk took place at the end of July, or even early in August. By reconstructing Jerne's research over the preceding years, Söderqvist has shown that, unlike Watson's insight, the selection theory was not closely connected to Jerne's ongoing research at the time. He regards the event as a true "conceptual leap," the other preconditions for which must be sought in a variety of background factors, ranging from Jerne's dislike of the prevailing template theory to a predilection for randomness and to his image of his own life.[16]

With the exception of an early student research project, Hans Krebs never mentioned in the accounts he later gave of his principal discoveries, nor acknowledged in my extensive conversations with him, having experienced a Eureka moment along the way to the solution of any of the problems he took up. That he should not have may appear puzzling, because the solutions that resulted in his two most important discoveries were of exactly the sort—noticing the way several pieces fit into a coherent pattern—that we might expect to occur to someone all at once.

During the weeks after he had observed the ornithine effect, Krebs tested various possible intermediates that might form a pathway connecting ornithine with urea, with negative results. During this period he entertained both the idea that ornithine was the source of the urea nitrogen and that its presence might somehow stimulate the formation of urea from ammonia. In January 1932 he began reducing the quantities of ornithine supplied to the tissue medium, until he reached such a low ratio of ornithine to urea formed that he could infer that the ornithine acts "like a catalyst." From this point onward he was, according to his later accounts, guided in his efforts to explain the ornithine effect by the "concept that a catalyst must take part in the reaction and form intermediates." He had in mind, however, no particular model of catalytic action. The other critical pieces of information he had were (1) that NH_3 is an obligatory intermediate, and (2) a reaction long known to take place in the liver by which arginine gives rise to ornithine:

$$C(=NH)(NH_2)NHCH_2CH_2CHNH_2COOH + H_2O = CH_2NH_2CH_2CH_2CHNH_2COOH$$
arginine   ornithine

$$+ NH_2C(=O)NH_2$$
urea

For some time, perhaps as long as a month, Krebs was unable to see how these pieces could be fit together, because in this reaction ornithine is simply the product of a process catalyzed by an enzyme. Eventually it occurred to him that the ornithine "may give rise to arginine." It could not do so by a reversal of the above reaction, because that would consume, rather than produce urea. When it became evident to him that there must be another route, he simply constructed the route in the most direct way possible, on paper, by combining ornithine with the two small molecules necessary to produce a balanced equation:

$$ornithine + 2NH_3 + CO_2 = arginine$$

(As mentioned previously, Krebs was immediately aware that this was a partial solution, because it required four molecules to combine at once. There must be at least two more, still unknown steps.) As soon as he had formulated this solution, Krebs found it compelling. "Everything fitted in," he remarked to me in 1976. It was, in other words, exactly the kind of solution that we might expect someone to see all at once. Yet Krebs associated the solution with no particular moment or background circumstance, and repeatedly described it as something that "became gradually clear."[17]

The solution that yielded the citric acid cycle similarly required Krebs simply to fit two or three pieces of a puzzle together. When he came across the paper by Martius and Knoop, he had already formulated and gathered experimental evidence for the existence in tissues of a reaction that would give rise to citric acid. In a lecture delivered the previous fall he had written, "Citric acid may be conceived as arising from malic and acetic acid by oxidative condensation. Malic and acetic acids, however, if added to tissues do not yield citric acid; pyruvic acid must be present." The pathway presented in the Martius-Knoop paper immediately completed the circle, leading from citric acid to α-ketoglutaric acid, which they themselves recognized "is then easily broken down through succinic, fumaric, and oxaloacetic acid to pyruvic acid, and is thereby connected to the oxidation products of the carbohydrates and several amino acids." By the time Krebs began the first experiments intended to verify that the pathway of Martius and Knoop takes place in surviving minced tissue, it appears evident from the circumstances that he must have fitted these pieces together into a cyclic pattern. The pattern was probably not identical to that which he submitted a few weeks later for publication, and may have included some ambiguity where malic, oxaloacetic, and pyruvic acid entered the cycle, but the general cyclic picture that emerged from fitting together these few pieces is, nevertheless, of a character one would expect to have appeared to Krebs at some holistic moment of recognition. He remembered no such event, although he agreed with the suggestion that at the time he began the experiments stimulated by the Martius and Knoop paper, "Maybe there was the hypothesis, yes. I think it is likely that the subsequent experiments were designed to test whether there was evidence for the cycle."[18]

In the case of the ornithine cycle, the temporal boundaries within which Krebs must have formulated the solution can be located only within a few weeks. In that of the citric acid cycle, the earliest possible date at which the Martius and Knoop paper could have reached the Sheffield library is close enough to that of the first recorded experiment clearly responding to that paper, that we can narrow the period to a few days. We can even conjure a highly plausible hypothetical scene in which Krebs is sitting in the library scanning the latest issue of *Hoppe-Seyler's Zeitschrift*, comes across the article, reads quickly through it, and arises a few minutes later with the first rudimentary images of a cyclic hypothesis already dancing through his mind, returns to his laboratory, and begins drawing arrows connecting the several sequences into a circular diagram.

Did Krebs have such moments of insight about which he subsequently forgot, or did he arrive at these solutions, as he remembered, by "hovering" over the problem until things gradually became clear? In the context of the solution of the ornithine cycle, he commented, "I don't know the details of the mental processes that make things gel, except that I do know it takes time."[19]

In contrast to such recollections by Krebs of gradual processes of mental clarification, Seymour Benzer remembered the unexpected conjunction of observations that gave rise to his rII mapping project as a powerful flash of insight. Here again I shall quote extensively from the account that Benzer gave in an essay written in 1965, about eleven years after the events it portrays. (For a summary of the first half of his account the reader may return to chapter 9, the paragraph ending at note 34.) After plating the mutant and wild type T2 phage on K12(λ) *E. coli* bacteria and finding again that the mutant gave no plaques, Benzer continued,

> To me the significance of this result was now obvious at once. Here was a system with the features needed for high genetic resolution. Mutants could be detected by the plaque morphology using strain B. Good high-titer stocks of the r mutants could be grown using strain K12S. Strain K12(λ) could be the selective host for detecting r⁺ recombinants arising in crosses between r mutants. A quick computation showed that if the phage genome were assumed to be on a long thread of DNA with a uniform probability of recombination per unit length, the resolving power would be sufficient to resolve mutations even if they were located at adjacent nucleotide sites. In other words, here was a system in which one could, as Delbrück later put it, "run the genetic map into the ground." I dropped everything else and embarked on the project.[20]

Thus, Benzer presented the coincidental conjunction of circumstances that gave rise to his insight as events happening within a relatively short time span, and the insight itself as occurring in a moment. While Benzer and I were examining his laboratory records, hoping to identify the point at which he had embarked on this project, he affirmed the character of his recollection by referring to what we were seeking as the "Eureka moment."

From the very complete collection of documents that Benzer has saved, it is possible to confirm nearly every element in the recollected series of coincidental events that he described as leading to the beginning of his project. Instead of taking place in a concentrated cluster as suggested in his account, however, they were spread out over a period of about fifteen months. He had given a genetics seminar on the "size of the gene"—which Benzer remembered as having prepared him to think about ways to establish that size by mapping (an event I have omitted from the summary in chapter 8)—on February 20, 1953, just before the publication of Watson and Crick's double helix, and, therefore, at a time in which he could not have imagined relating the distance between mutations to the dimensions of nucleotides. I have not yet found out when Benzer may have prepared the stocks of r mutant phage for the Hershey-Chase experiment, or plated T2r and T2r$^+$ on lysogenic and non-lysogenic bacterial strains, but the date at which he began systematic research on the phage mutants is clearly recorded, at the beginning of a notebook entitled "r mutants," as January 9, 1954. The experiments continue in an unbroken series until Benzer left for Cold Spring Harbor at the end of June. They do not show, however, that he "dropped everything else" to embark on the mapping project. Although he did perform some crosses on r mutants during the first two weeks, for the next three months he utilized the r mutants for different purposes. Testing them on various other lysogenic and non-lysogenic strains, he worked on a problem that he described in a letter written on March 2 as "the phenotypic expression of the r mutant," and in another letter of the same day as "a problem of interference between carried and infective bacterial viruses." Not until early May does the laboratory record show Benzer beginning the mapping project in earnest. After that, he pursued it single-mindedly through May and June. On May 28 he wrote down a page of "thoughts on the gene" that suggests he was now beginning to think about the relation between recombination frequencies and the number of "nucleotides per unit of recombination," and it was probably only sometime after that that he made what he referred to in his recollection as the "quick computation" showing that the resolving power of his system would enable him to resolve mutations no more than a nucleotide apart.[21]

How can we explain the manner in which Benzer's memory compressed events that had occupied more than a year into a single, concentrated "Eureka experience?" According to Gruber's analysis, this is a common phenomenon, not unlike Darwin's compression of a series of weaker encounters with ideas like natural selection into the single, powerful Malthus experience. Ever simplifying, memory regularly suppresses those aspects of a complex progression of events that did not prove, in retrospect, essential to an outcome. Points along such progressions that are accompanied by strong emotions, such as the excitement one may feel when first recognizing a new possibility, are apt to overshadow what came before and after under calmer circumstances, just as in Gruber's metaphor the crest of a wave more strongly impresses us than the phases that precede or follow it. We need not doubt that Benzer actually experienced a forceful moment at which the first intimations came to him that he had at hand a potential system "with the features needed for high genetic resolution," merely because in the intervening years he had forgotten that he did not drop everything else as quickly, or follow up on his insight as directly, as he later thought.

When we compare the case of Hans Krebs, who remembered as gradual clarifications the solutions of problems that appear by their nature more likely to have come to mind all at once, with that of Benzer, who remembered as a compact, if not instantaneous, event a series of developments lasting more than a year, we may come to question the significance of the very concept of a Eureka moment. Are these experiences illusions, epiphenomena that may have subjective impact but little to do with the mental processes by which scientists actually reach solutions? According to some cognitive scientists, creative thinkers have no direct access to their own cognitive processes. On the other hand, it is hard to deny the subjective reality of the experience. Most of us who have thought hard and long about anything can testify that we have more than once had the feeling that the answer to a question with which we have been occupied seems somehow to emerge into our awareness in a moment of insight.

Gruber's interpretation of the "Aha experience" seems to me to show the way around this dilemma. He wrote:

> There are two seemingly opposite approaches to creative work. One emphasizes sudden moments of insight, dramatic reorganizations of ideas. . . . The other emphasizes the slow construction of ideas, treating creative thinking as a growth process. Should we think of the "act" or the "process" of creation?
>
>   Three different approaches to this question come to mind: (a) these two ideas are opposed and mutually exclusive; (b) they simply represent different poles of a continuous spectrum of possibilities; (c) they are complementary ideas and we need to consider in some detail the manner in which they articulate with one another.[22]

Although Gruber believed himself to be exploring primarily the third of these alternatives, his analysis seems to support both of the last two. His view that Eureka moments are nodal points in a continuous growth process develops the idea that moments of insight and slow construction articulate with each other. His explication of the structure and temporal duration of a Eureka experience implies, however, the spectrum approach. If such an experience is not truly instantaneous, and the lower limit of a process in which a significant insight can emerge is a few seconds, there appears to be no upper limit, except for the loss, at some degree of extension, of the subjective feeling that the insight has emerged as a whole. Moreover, the process of acquisition of an insight that feels like an extended episode while it is happening may, with the passage of time, come to be remembered as unitary.[23]

As my examples imply, we can probably never have a record of the thoughts of an individual surrounding such an experience preserved in such minute detail as to permit us to isolate the boundaries of any particular historical insight occurring on the way to a scientific achievement, or the immediate mental transitions through which such insights may be articulated with slower growth processes. By recognizing, however, the lack of a sharp opposition between these two patterns of thought, we assure that our inability to do so does not defeat our endeavor to understand the fine structure of investigative pathways.

Conclusion: The Narrative Representation of Investigative Pathways

In the preceding chapters we have viewed aspects of the individual investigative pathway at various levels, beginning with broad phases in the scientific life, moving downward to smaller-scale episodic rhythms, and finally approaching the fine structure of daily activity. We may now ask how may one go about incorporating the patterns observed at these several scales into a historical narrative?

Robert Richards has expressed the view that a narrative is inevitably teleological. A historian writes "toward an ending," and selects the events and details to be included with foreknowledge of how they will contribute toward, or pose obstacles to, the denouements in the story.[1] If we accept the view of François Jacob that the scientists whose investigations we portray can never know in advance how things will turn out, can the historian who *does* know how they turned out recapture the openness of the possibilities that had existed before the story, or before the episodes within the story have reached their denouements? If we accept, with Rheinberger, that after a surprising event has occurred, it "becomes more and more difficult, even for the participants, to avoid the illusion that it is the inevitable product of a logical inquiry or of a teleology of the experimental process,"[2] how can the historian avoid portraying what has already happened by the time she writes about it as the inevitable outcome of the investigative pathway leading toward it?

The assertion that an outcome is not inevitable implies that alternative outcomes were possible, had some contingent circumstance along the way not taken place, or taken place differently. All causal interpretations of historical events imply counterfactual possibilities, other courses that might have been followed had particular factors been added or removed. We know also, however, that it is impossible to pursue such counterfactual scenarios beyond vague hints, because the number of possibilities quickly becomes divergent.

These dilemmas, inherent in any historical narrative, are further complicated in the history of science, because the possibility of alternative outcomes immediately raises the tangled question of whether scientific knowledge is itself contingent and conditioned by extrinsic factors, or whether it is hidden within "nature" until it is revealed by appropriately conducted investigations. Historians committed to one or the other of these positions will undoubtedly shape their narratives to support whichever side they wish to illustrate. Fortunately, however, we need not resolve these issues if we restrict ourselves to recovering how our subjects thought and acted, without claiming a transcendent understanding of the status of their achievements or failures.

If we cannot avoid teleological inevitability by exploring other hypothetical outcomes, we can do so by including as full accounts as we can of the directions our subjects took that seemed for a time to point toward other outcomes that were not realized. Finely detailed narratives allow room for the openness to recover all those possibilities that our subjects thought might come to pass, without giving away in advance which of them, or even if an entirely unanticipated one, eventually did emerge. It is when we compress and simplify our stories that we are forced, in order to retain all those steps that did prove necessary to the outcome, to eliminate those elements that did not, in the end, prove essential. The more we do so, the more we approach the illusion that the entire process followed a parsimonious, logical course toward an inevitable conclusion.

We cannot easily escape from our dilemma by resisting simplification, however, because the full story quickly becomes too complex to reconstruct in a single narrative. Selection is inevitable. In the foregoing chapters I have illustrated the various dimensions at which the investigative pathway of an individual can be followed. How can these be combined into a single narrative? In my study of Hans Krebs I felt that I had reached the limit of such a construction in following him day by day through the first decade of his career. Some readers believe that I far exceeded such limits. If, however, we cannot sustain an unbroken pathway over the lifetime of our subjects, how can we proceed? Gruber suggests, in his study of Darwin, that it is necessary to go back over the same episodes at various levels of resolution, to bring out different dimensions of creative activity. As in many other studies of complex phenomena, our own cognitive limitations force us to compromise, to give up at some point complete description, and to rely on sampling techniques to sketch what we cannot fully portray.

That is not necessarily a counsel of despair. In every realm of our lives we confront complexity beyond our capacity fully to comprehend, and we are forced to find ways to simplify what we encounter. If we can only seldom reconstruct every step that a scientist takes along her way, and only then for limited stretches, we can still return to the pathway metaphor to understand how to fit such stretches into broader images of the investigative process and the scientific life, and to give them all deeper meaning.

Afterword: F. L. Holmes and the History of Science

Jed Z. Buchwald

The most cursory reader of contemporary history of science will rapidly discover the work of Larry Holmes. He has set standards, and offered models, that other historians can only hope to emulate, both of which are amply evident in this book, *Investigative Pathways*. In so doing Larry has uncovered structures in scientific practice that exemplify what might best be termed the *Holmsian pattern*. In this afterword I will draw on areas distant from those which Larry has himself studied in order to show that this is very much a general pattern, and one, moreover, that points historians who aim to uncover the complex routes followed by scientists, as well as the deep practices that underlie their research, in just the right directions.

We begin with one of Larry's many salient points, namely that a community's apparently surprising support for what might appear in hindsight to be novel and even unorthodox work frequently derives neither from the community's inability to perceive a possible challenge nor from its desire to support novelty. It may rather have much to do with the methodological canons that the new work embodies. In particular, Holmes's Lavoisier received initial support precisely because his work embodied rigorous quantitative methods—marking Lavoisier as "a very able practitioner" of contemporary chemistry. Very much the same reaction greeted another Frenchman a generation later, and at the hands of one of Lavoisier's own collaborators, namely Pierre Simon de Laplace.

Many historians and scientists have over the last two centuries remarked with some surprise that in 1818 the young Augustin Fresnel won a prize competition on the subject of the diffraction of light when the judging commissioners included both Laplace and Poisson, both of whom had very different views, to say the least, about the nature of light than the one that Fresnel now advanced.[1] Why did they agree to award him the prize, despite this difference and the fact that Laplace told Fresnel that he was troubled by the lack of simplicity in Fresnel's mathematics and that he could not accept Fresnel's physical assumptions? One explanation of this apparent conundrum refers to Poisson's having deduced an apparently unintuitive consequence from Fresnel's mathematics, which was rapidly found to hold in the laboratory. The confirmation of Poisson's deduction is certainly a significant fact—not so much for the experimental success per se, but rather for its congruence with what truly did impress the commissioners, namely, that Fresnel's experiments and calculations superbly exemplified the combination of exact and careful measurement with computational structure that by this time had become the sine qua non for success in Parisian scientific circles. Indeed those who were not able to satisfy these criteria rapidly found themselves with little respect among their contemporaries. Ironically, among those who were not able to meet contemporary standards was Fresnel's own patron, François Arago—and it was in part at least because Arago understood very well just what he lacked, but that the young Fresnel could provide, that he so strongly supported Fresnel.

Neither Larry's Lavoisier nor the young Fresnel exemplifies isolated or unusual moments in science history. Indeed, as Larry so thoroughly shows, those who have mastered prevailing standards of scientific practice stand a good chance of gaining a sympathetic and attentive hearing even for radically novel work. The outsider who has a new idea but who has little mastery of the tools of the trade will be thought of as a poor craftsman and will not easily gain entry to the support mechanisms of a field.

But more than adept practice may usually be necessary to gain appropriate entrée, and here too Larry has emphasized something that is frequently quite important, namely how it happens that someone gains enough access to the organs and support mechanisms of a field to have work pushed forward and given a sympathetic hearing. Larry has discussed five cases in each of which the scientists, in his words, "reached positions of early distinction in part through some form of privileged access." There are many forms of privileged access—Larry has discussed several—but among them are access to community members who for one reason or another will push forward and support the neophyte. I will single out two examples of this that fit Larry's pattern: Fresnel again, and then the young Heinrich Hertz, over half a century later in Germany.

Fresnel gained support and entrée among the elite of Parisian science for three reasons. First, he had the requisite training, which he acquired at the Ecole Poly-technique. This was more a necessary than a sufficient condition for success, however. Second, he had a patron in the person of Arago, whose ascension to a position of growing power had more to do with the vagaries of the early Napoleonic wars than with Arago's own expertise in contemporary practice. (Arago had returned to Paris a hero after having been thought lost.) But how did Fresnel, third, gain Arago's attention? He was able to do so not least because Arago was friendly with Fresnel's maternal uncle, Léonor Merimée. Appropri-ate education and social connections accordingly provided Fresnel with the kind of access that gave him the resources to create and successfully to publicize the novel science of wave optics within less than a decade.

Heinrich Hertz provides a second apposite example.[2] He was patronized (probably in both senses) by the immensely powerful and influential Hermann von Helmholtz in late 1870s and early 1880s Berlin. Hertz had the right back-ground and contacts for Helmholtz to have taken him seriously in the first place, but the contours of Hertz's own practice, and the professional support that he received from Helmholtz, were what shaped the early outlines of his tragically short career. For Hertz took as his own the deepest reaches of Helmholtz's way of doing physics, and even many of the specific problems with which Helmholtz was grappling during the 1870s in electrodynamics. And it was Helmholtz who employed Hertz as a research assistant and who then wrote recommendations that helped land Hertz his first position in Kiel—even though Helmholtz likely did not want Hertz to accept. Further, Hertz's displacement from Berlin, to-gether with the comparative aloofness of his mentor, corresponds in time with the first indications of Hertz's moving intellectually and in the laboratory in di-rections that soon became uniquely his own. Here, in Hertz's attitudes toward his mentor, we can see the complicated ways in which access to and interaction with those who hold the reins of power in a discipline can markedly affect the content and course of an individual's research.

Larry's insights into the workings of scientific research have much to say about issues concerning competition between research programs as well as between individuals. There is a canonical image of competition that continues to inform popular imagery, namely that a new system comes along that conflicts with the old one, which has been around for some time, with battles resulting. Though it is no doubt reasonable to say that most historians of science nowadays would not accept quite so simple an image as this, nevertheless essential aspects of the picture continue to be influential. In particular, many would probably agree that there is a well-defined temporal sequence in most cases of this sort, such that a preceding, and stable, system generally does confront a new one that must accommodate aspects of the old which have stood the test of time.

Larry tells a rather different story. He points out, in the case of Lavoisier, that nothing quite like this ever did take place, because Lavoisier's and Priestley's systems, which have respectively occupied the places of new and old in most histories, were effectively contemporaneous. There was nothing like a stable scheme in relevant areas for Lavoisier to react against. In Larry's words, "the contest between Lavoisier and Priestley that shaped much of the dynamic of the chemical revolution can be viewed as a competition between two contemporary research programs, rather than the overthrow of an established paradigm." This is a critically important insight that deserves to be thoroughly explored, and I'd like to further it by drawing again, briefly, on the history of optics.

Most scientists and historians of science continue to think that there was a battle, indeed a very public battle, between two optical systems, the one much younger than the other, that eventuated in the triumph of wave optics by the 1830s. It is certainly true that wave optics had become the predominant form of optical science by that time; and it is also true that most of the instruments that pertain to wave optics, and indeed much of what wave practitioners did on paper and in the laboratory, had no precedent in the eighteenth century. Nevertheless, the spread of wave optics is in fact a very good example of exactly the kind of situation that Larry had in mind, and that is probably quite general.

For the fact is that there was *no* preexisting alternative to wave optics that could do the kinds of things that wave practitioners were particularly concerned with until the years during which Fresnel produced his novel system. That is, there was no older system with which wave optics partisans actually had to contend. This is a rather surprising claim considering most historiography on the subject, but then so was Larry's in respect to Lavoisier and Priestley, for many of the same reasons.

What I have in mind is this. Until the first decade of the nineteenth century, practical optics—both in theory and in the laboratory—was for the most part a science of light rays: of objects (or the paths of objects) that were considered to exist as individuals and that could be refracted, reflected, and even bent (in this last case explaining the existence of diffraction, or what Newton had termed inflection). There was, moreover, a mathematical science of these things—namely, geometric optics—and this science persisted in very much the same form after the production of wave optics, with the new (and certainly novel) proviso that the dimensions of optical instruments had to be taken into account in deciding whether geometric principles could be applied. This latter proviso was certainly difficult—Poisson, for one, had trouble with it—but it is not here that we find contention. Not at all: on these grounds acceptance was rapid and, if not altogether without controversy, nevertheless quite smooth. There was in fact little to contend about just because wave optics could actually calculate things in respect to diffraction that no one, including Newton, had ever tried to do for rays. There was, in other words, no preexisting, developed system for the quantitative, experimentally grounded handling of a phenomenon (inflection) that had previously not seemed to be very important—a phenomenon, moreover, that, from the point of view of people like Laplace, remained rather marginal.

Yet there certainly was a great deal of contention over wave optics. It's just that the battle did not occur over something that required rejecting a long-standing, highly stable practice. Not at all, for the most intense battles occurred over issues that involved quite recently discovered phenomena which French scientists had been highly successful in handling in the laboratory, and for which there was a developing quantitative system: phenomena associated with the polarization of light. Here, starting with the discovery of polarization's general properties by the Polytechnician Etienne Louis Malus, a veritable swarm of novel discoveries were pouring forth from French, English, and Scottish laboratories at almost the very moment that Fresnel was himself brought to Paris by Arago to abort the process.

Fresnel's most deeply creative work, in both theory and experiment, occurred as he grappled with the issues posed by polarization. During the very years that he did so, scientists in France, England, and Scotland were themselves developing highly successful ways to handle the new discoveries that they were making, ways based directly on the redistribution and counting up of sets of light rays. Yet these very procedures were forbidden by Fresnel's wave optics, which was being created at nearly the same moment in time. Here, therefore, we do have a deep and engaging problem of accommodation and rejection between very different systems, with the signal difference from popular myth that these alternatives were born, developed, and engaged each other agonistically all during the same period. There was no ancestral father-figure that had Oedipally to be rejected by wave partisans, though there was something rather like sibling rivalry.

Larry's chemical revolution and the one I've briefly touched on in optics exhibit quite similar patterns, and not just because both occurred in France, the one about a generation after the other, separated from it by the great divide of the Revolution. There are other examples from different times and different places. The production of field theory by Faraday, Kelvin, and Maxwell is one such—since the major contending system to field theory, namely the German physicist Wilhelm Weber's electric particles, emerged at just about the same time. Larry's lesson here is a quite general one, and it has important implications as well for the philosophy of science.

Larry's work contains many other lessons, and primary among them has been his careful, detailed excavation of all the available evidence in order to reconstruct the changing meanings and complex pathways that produce novel results. His work here is particularly significant as an exemplar of what can and should be done by historians who engage with the actual content of scientific work. His analysis of Krebs's route to the eponymous cycle stands as a beautiful instance of what to do in this vein. Reading it, as well as Larry's other studies of the kind, we are virtually present with the investigator through the pathways, byways, serendipitous insights, and experimental surprises, that always mark truly significant discoveries.

I'd again like to draw on an example from optics that fits the pattern Larry has delineated. Shortly before 1820, Fresnel was faced with the most daunting task of his career, which was to produce a wave theory for the optical behavior of crystals. As always he turned to a complex, and complexly evolving, combination of theory, calculation, and experiment to probe this difficult, recalcitrant area—the very one that had been the greatest, and recent, triumph of his compatriots.

In the 1810s, David Brewster in Scotland, Jean-Baptiste Biot in France, and John Herschel in England, among others, had explored the colored patterns that are generated when polarized light passes first through a thin crystal and then through another polarizer. Among other things, they knew that crystals come in two classes in respect to this phenomenon of chromatic polarization, which are differentiated from each other according to whether, under appropriate circumstances, one or two sets of colored rings are produced. The single-ring class included crystals like calcite; in these crystals one has a single optic axis, with an entering light ray splitting into an ordinary and an extraordinary refraction, the latter of which is governed by the incident ray's angle with respect to this optic axis. Brewster et al. reasoned that crystals in the other class have two optic axes, with each axis behaving in the usual way in respect to light, thereby resulting in three refractions in all within these kinds of crystals: one ordinary and two extraordinary.

Fresnel, on the basis of wave principles, early conjectured that in fact there must be just two refractions, not three, and that neither of these two could behave like the old ordinary ray, that is, that neither one could obey Snel's law. This was a highly disruptive claim in the contemporary context, and in a brilliant series of experiments Fresnel was able to confirm it to his and soon to others' satisfaction, despite the novelty (because, again, Fresnel's work beautifully exemplified the investigative canons of contemporary Parisian science). But that was not all, and what Fresnel subsequently did provides a very nice illustration of the unexpected byways of experimental investigation that Larry has strikingly pointed out in so many instances.

Fresnel wished to produce an account that would generate a complete way to calculate the behavior of these biaxial crystals. This meant that in the end he wanted to generate for them the analog of what is known as the wave surface in uniaxial crystals such as calcite. But Fresnel also needed to link this as yet unknown wave surface to polarization properties, so that both could be deduced together from a single structure, and this led him into a complicated, evolving mix of conjecture with experiment. The first product of these conjectural probes was an ellipsoid that was not the wave surface proper, but rather one that could be used to calculate the wave surface as well as polarizations.

I come now to the point. Fresnel probed this surface in the laboratory in extraordinarily clever ways, and he concluded that it works extremely well. There were a few discrepant observations, and these he at first attributed to experimental error. He remarked at the time that he regarded these attempts as only a "first, approximate verification of the theory." He was quite right at this point to attribute the differences he had found to experimental error, but the ironic fact is that within a very few weeks Fresnel had changed his mind and had decided that the problem lay rather with theory than with observation. What he realized was that he, Fresnel, had made the elementary mistake, in this new optics of waves, of conflating wave speeds with ray speeds, whereas the two are generally quite different things in crystals. This meant that the ellipsoid in question should not in fact have worked well, and so Fresnel shifted the error off the back of apparatus and onto that of theory, where he sought to make corrections. This search, propelled by a combination of measurement deviation with theoretical realization, eventually did produce a way to calculate wave (not ray) speeds, and from this the wave surface could itself be found.

But we are not done (though almost). Fresnel could not in fact figure out how to do the calculation; it's quite complicated and was first accomplished later by his friend, André-Marie Ampère of electrodynamics fame. Nevertheless Fresnel did produce the wave surface, the very one that we still use today. How did he do so? By using the very first ellipsoid, based on ray speeds, whose theoretical impropriety had led him into a new search several months before. Reasoning backward from the behavior of uniaxial crystals, Fresnel was able to generate the analog of his earlier ellipsoid for them, and this, when generalized, produced the very same ellipsoid that he had previously concluded could not be correct, but that he now decided must work well after all. In which case he had been right in the first place to attribute the deviation between it and experiment to measurement error and not to theory, even though the way he had originally obtained the now-acceptable surface was in the end hardly acceptable at all.

Here then we have a most interesting case of an investigator moving backwards, forwards, and even sideways between conjectural calculation and complicated experiments in an effort to reach a persuasive result. The route was not at all straightforward, neither was it a case of theory simply driving experiment or the other way around. Each propelled the other, and both were driven by Fresnel's desire altogether to replace the earlier claims of Brewster and Biot, thereby thoroughly cementing the novelty of wave optics. This is yet another illustration of the curvy investigative pathways that Larry has brought out more clearly than anyone else.

We turn now to the overall character of Larry's work, and why it is more important than ever that his approach to history, his methods and goals, should influence the way that the history of science is practiced, since even a cursory glance at the contents of many contemporary journals in the field, and most monographs, will show that what most concerns Larry, and what should most concern the field, does not occupy its center.

More deeply and with greater insight than other historians of science, Larry evokes the actual practice of scientific investigation. Of course there are many examples one could point to today of historians who also claim to do just that. But what we will often find are displays of what one might, no doubt controversially, call the detritus of scientific work. We may find many pictures of apparatus, of people working at tables, of (for modern science) minutes of meetings, of computer printouts, or even of emails. And we may even find brief précis, as it were, of what people were doing. All are, or can be, quite illuminating. But the focus of the resulting account will not usually be on the intellectually intricate and practically complex production of novel scientific research to which this detritus bears witness. Indeed, some modern histories of science would actually reject the very meaning of trying to do such a thing. But that is precisely what Larry has done, and it should lie at the very core of history of science.

Larry's several works—on Claude Bernard, on Lavoisier, Krebs, Meselson and Stahl, as well as his current research on Seymour Benzer—probe thoroughly the complex ways in which novelties emerge in the laboratory. Here we find neither simplistic stories of success or failure based on ex post facto knowledge, nor the equally simplistic reduction of discovery to social manipulation or cultural happenstance. This is not at all to say that Larry's work eschews sociocultural factors. Not at all—but their presence in his narratives is in the service of understanding the intricate manner in which scientific work is produced: they are not in themselves the substance or aim of his stories. As Larry's colleague at Yale, John Warner, has remarked, Larry binds the lives and idiosyncrasies of his subjects closely to his accounts of their laboratory work.

A passage from Larry's *Lavoisier and the Chemistry of Life* exemplifies the judicious combination of judgment, understanding, analysis, and historical insight that characterizes Larry's work. In discussing difficult experiments undertaken by Laplace and Lavoisier on heats of combustion in two circumstances that they thought should be one and the same, and that other historians had gone over unconvincingly beforehand, Larry writes:

> Given the uncertainties of the experimental data, Lavoisier and Laplace were justified in concluding that the agreement [between measurement and calculation] was sufficiently close to support strongly their view that these were measurements of the heat of the same combustion process. On the other hand, we can see that the degree of agreement they achieved depended partly upon some further assumptions that were themselves uncertain, and that by making other plausible assumptions one could easily reach a result which would appear far less favorable to their conclusion. What they had been able to provide was a result compatible with a theory they had previously embraced, that reinforced their confidence in its correctness, and that added a new dimension in which respiration could be quantitatively compared with combustion. In the long run what was most important about these experiments was not the agreement between the two results, but that they had demonstrated the feasibility of measuring the quantities of the heat released in both processes. In 1777 Lavoisier had doubted whether such measurements could be made.[3]

Here we see how Larry combines a careful, critical understanding of what his subjects were consciously up to, with thorough knowledge of presumptions that they themselves did not explicitly articulate, to arrive in the end at a nuanced view of what the contemporary consequences of the investigations turned out to be. This is not the mark of a historian afraid to judge; it is the mark of one who judges through deep analysis neither retrospectively to criticize nor to celebrate, but rather to understand.

Understanding as a goal marks all of Larry's work in the history of science. It is hardly a widespread one nowadays. There is a great deal of good and very interesting work being produced in science history, perhaps, in some ways, more than there has ever been before. But very little of it is concerned with the kind of deep and sympathetic understanding of the content of scientific practice that marks Larry's work, and that he has explicitly articulated in his *Investigative Pathways*. Neither careful excavation of the byways and pathways of experimental or theoretical work, nor the provision of careful insights into the individual's practice as a member of an investigative community is nowadays common among historians of science. They do mark all of Larry's work. Indeed, one historian of science has characterized Larry's penetrating insights into eighteenth-century chemistry, wherein he tried to excavate the enterprise from within its own structure insofar as possible, as "heroic anti-Whiggery," which just shows how rare the attempt actually is.

It is fitting to conclude with another quotation from Larry's work, this one drawn from the last page of his "white book" on eighteenth-century chemistry. He writes:

> The historical study of science as an investigative enterprise incorporates ideas, but does not treat them as having a life of their own. It shows how ideas come into and emerge from sustained explorations of aspects of nature that can be confronted in the laboratory, the observatory, the museum, or the field. It is attentive to the ways in which scientists organize themselves, share power and authority, promote their interests, and award or withhold recognition; but it does not forget that scientists do all these things in the first place in order to facilitate the investigative activity around which they orient their lives.[4]

History of science that aims to provide deep insight will carry out Larry's investigative enterprise.

Notes

INTRODUCTION

1. Dean Keith Simonton, *Origins of Genius: Darwinian Perspectives on Creativity* (New York: Oxford University Press, 1999), pp. 145–197.
2. Howard E. Gruber, *Darwin on Man: A Psychological Study of Scientific Creativity,* 2d ed. (Chicago: University of Chicago Press, 1981), pp. 6, 113.
3. John Ziman, *Knowing Everything about Nothing: Specialization and Change in Research Careers* (Cambridge: Cambridge University Press, 1987), pp. 1–21.
4. Russell Kahl, ed., *Selected Writings of Hermann von Helmholtz* (Middletown, Conn.: Wesleyan University Press, 1971), p. 474.
5. Claude Bernard, *Cahier de notes,* ed. M. D. Grmek (Paris: Gallimard, 1965), pp. 128–129.
6. Vincent du Vigneaud, *A Trail of Research in Sulfur Chemistry and Metabolism and Related Fields* (Ithaca: Cornell University Press, 1952), p. vii. I thank Joseph S. Fruton for bringing this reference to my attention.
7. Ziman, *Knowing Everything,* p. 20.
8. See George Lakoff and Mark Johnson, *Metaphors We Live By* (Chicago: University of Chicago Press, 1980).
9. Ziman, *Knowing Everything,* p. 1.
10. K. Anders Ericsson, "The Acquisition of Expert Performance: An Introduction to Some of the Issues," in *The Road to Excellence,* ed. K. Anders Ericsson (Mahwah, N.J.: Lawrence Erlbaum, 1996), pp. 10–13.

11. Michael Polanyi, *Personal Knowledge: Towards a Post-Critical Philosophy* (Chicago: University of Chicago Press, 1962), p. 61.

12. On the general argument that what count as discoveries are events on which that status is conferred by "members of society," see Augustine Brannigan, *The Social Basis of Scientific Discoveries* (Cambridge: Cambridge University Press, 1981), esp. p. 64.

CHAPTER 1. INVESTIGATION, DISCOVERY, AND EXPERIMENTAL PRACTICE

1. Robert Frank, *Harvey and the Oxford Physiologists: A Study of Scientific Ideas* (Berkeley: University of California Press, 1980), pp. xi–xii.

2. Martin Rudwick, *The Great Devonian Controversy: The Shaping of Scientific Knowledge among Gentlemanly Specialists* (Chicago: University of Chicago Press, 1985), quotation p. 12.

3. Thomas Nickles, "Introductory Essay: Scientific Discovery and the Future of Philosophy of Science," in T. Nickles, ed. *Scientific Discovery, Logic, and Rationality* (Dordrecht: D. Reidel, 1980), p. 2.

4. Kenneth F. Schaffner, *Discovery and Explanation in Biology and Medicine* (Chicago: University of Chicago Press, 1993), pp. 8–63, quotation p. 8.

5. Lindley Darden, *Theory Change in Science: Strategies from Mendelian Genetics* (New York: Oxford University Press, 1991).

6. David Gooding, *Experiment and the Making of Meaning: Human Agency in Scientific Observation and Experiment* (Dordrecht: Kluwer, 1990), entire, quotations pp. 15, 18.

7. Andrew Pickering, *The Mangle of Practice: Time, Agency, and Science* (Chicago: University of Chicago Press, 1995), p. 2.

8. Barry Barnes, *T. S. Kuhn and Social Science* (London: Macmillan, 1982); Jan Golinski, *Making Natural Knowledge: Constructivism and the History of Science* (Cambridge: Cambridge University Press, 1998), pp. 13–26.

9. Bruno Latour and Steve Woolgar, *Laboratory Life: The Social Construction of Scientific Facts* (Beverly Hills: Sage, 1979), pp. 238, 243.

10. Steven Shapin and Simon Schaffer, *Leviathan and the Air-Pump: Hobbes, Boyle, and the Experimental Life* (Princeton: Princeton University Press, 1985).

11. Steven Shapin, "History of Science and Its Sociological Reconstructions," *History of Science* 20 (1982): 157–211, quotations pp. 164, 194.

12. For a lucid summary and sympathetic appraisal of the constructivist view, see Golinski, *Making Natural Knowledge.*

13. Pickering, *Mangle*, p. 1.

14. Pickering, *Mangle*, entire.

15. Pickering, *Mangle*, p. 3

16. Jed Z. Buchwald, *The Creation of Scientific Effects: Heinrich Hertz and Electric Waves* (Chicago: University of Chicago Press, 1994).

17. J. L. Heilbron, *Weighing Imponderables and Other Quantitative Science around 1800* (Berkeley: University of California Press, 1993); Alan J. Rocke, *The Quiet Revolution: Hermann Kolbe and the Science of Organic Chemistry* (Berkeley: University of California Press, 1993); Mary Jo Nye, *From Philosophical to Theoretical Chemistry: Dynamics of Matter and Dynamics of Disciplines, 1800–1950* (Berkeley: University of California Press, 1993); Ger-

ald L. Geison, *The Private Science of Louis Pasteur* (Princeton: Princeton University Press, 1995).

18. Angela N. H. Creager, *The Life of a Virus: Tobacco Mosaic Virus as an Experimental Model, 1930–1965* (Chicago: University of Chicago Press, 2002), quotation, p. 4; Nicolas Rasmussen, *Picture Control: The Electron Microscope and the Transformation of Biology in America, 1940–1960* (Stanford: Stanford University Press, 1997); Nathaniel C. Comfort, *The Tangled Field: Barbara McClintock's Search for the Patterns of Genetic Control* (Cambridge: Harvard University Press, 2001).

19. Daniel P. Todes, *Pavlov's Physiology Factory: Experiment, Interpretation, Laboratory Enterprise* (Baltimore: Johns Hopkins University Press, 2002).

20. Frederic L. Holmes, Jürgen Renn, and Hans-Jörg Rheinberger, eds., *Reworking the Bench* (Dordrecht: Kluwer, 2003).

CHAPTER 2. THREE SCALES OF THE INVESTIGATIVE PATHWAY

1. Fernand Braudel, "La longue durée," in *Ecrits sur l'histoire* (Paris: Flammarion, 1969), pp. 41–56.

2. Ibid., p. 55.

3. Howard Gruber, *Darwin on Man: A Psychological Study of Scientific Creativity,* 2d ed. (Chicago: University of Chicago Press, 1981), pp. xxi, 5, 10, 114.

4. Martin Rudwick, *The Great Devonian Controversy: The Shaping of Scientific Knowledge among Gentlemanly Specialists* (Chicago: University of Chicago Press, 1985), pp. 7–8.

5. *Faraday's Diary: Being the Various Philosophical Notes of Experimental Investigation Made by Michael Faraday,* ed. Thomas Martin, 7 vols. (London, 1932–1936).

6. Friedrich Steinle, "The Practice of Studying Practice: Analyzing Laboratory Records of Ampère and Faraday," in *Reworking the Bench,* ed. Frederic L. Holmes, Jürgen Renn, and Hans-Jörg Rheinburger (Lancaster: Kluwer, 2002), pp. 93–117.

7. For the outstanding example of Charles Darwin's voluminous correspondence, see Janet Browne, *Charles Darwin: The Power of Place* (New York: Knopf, 2002), pp. 12–13.

8. Rudwick, *Devonian Controversy,* p. 8.

9. The complete correspondence of Liebig and Wöhler is contained in Liebigiana, Handschriften Abteilung, Bayersche Staatsbibliothek, Munich. The nineteenth-century published edition, *Wöhler und Liebig Briefe von 1829–1873,* ed. A. W. von Hofmann and reissued by W. Lewicki (Göttingen: Cromm Verlag, 1982), is inaccurate and incomplete.

10. Max Delbrück Archive, Archives of the California Institute of Technology.

11. François Jacob, *Of Flies, Mice, and Men* (Cambridge: Harvard University Press, 1998), pp. 12–13.

12. Peter Galison, *How Experiments End* (Chicago: University of Chicago Press, 1987), p. x.

13. [Bourdelin], "Reçueil d'analyse chimique," unpublished laboratory notebooks, 11 vols., Bibliothèque Nationale, Paris.

14. See Frederic L. Holmes, "The Communal Context for Etienne-François Geoffroy's *Table des Rapports,*" *Science in Context* 9 (1996): 289–311.

15. Hans Sloane Collection, Archives of the British Library, London.

16. "Cahiers de laboratoire de Lavoisier," Archives of the Académie des Sciences, Paris.

17. See Mirko D. Grmek, *Catalogue des manuscrits de Claude Bernard* (Paris: Masson, 1967).

18. Deposited in the "Krebs Collection," Sheffield University Library.

19. Lorraine Daston, *Wunder, Beweise und Tatsachen: Zur Geschichte der Rationalität* (Frankfort am Main: Fischer, 2001), pp. 1–27.

CHAPTER 3. APPRENTICESHIP AND INDEPENDENCE

1. Hans Krebs, "The Making of a Scientist," *Nature* 215 (1967): 1441–1445.

2. John E. Lesch, *Science and Medicine in France: The Emergence of Experimental Physiology, 1780–1855* (Cambridge: Harvard University Press, 1984).

3. Frederic L. Holmes, *Claude Bernard and Animal Chemistry* (Cambridge: Harvard University Press, 1974), pp. 127–131.

4. Frederic L. Holmes, *Hans Krebs: The Formation of a Scientific Life, 1900–1933* (New York: Oxford University Press, 1991), p. 133.

5. Krebs, "Making of a Scientist," p. 1443.

6. Ibid., p. 1442.

7. Ibid.

8. M. D. Grmek, *Raisonnement expérimental et recherches toxicologiques chez Claude Bernard* (Geneva: Droz, 1973), pp. 22, 63.

9. Ibid., passim.

10. Jed Z. Buchwald, *The Creation of Scientific Effects: Heinrich Hertz and Electric Waves* (Chicago: University of Chicago Press, 1994), pp. 59–63.

11. Ibid., pp. 75–76, 99, 170–171, 173.

12. Ibid., 324, 325.

13. For a meticulous effort to track the later lives of both eminent and little-known figures trained under influential leaders of several chemical and biochemical laboratories during the nineteenth and early twentieth century, see Joseph S. Fruton, *Contrasts in Scientific Style: Research Groups in the Chemical and Biochemical Sciences* (Philadelphia: American Philosophical Society, 1990).

14. Henry Guerlac, *Antoine-Laurent Lavoisier: Chemist and Revolutionary* (New York: Scribner's, 1975), pp. 51–53.

15. Arthur Donovan, *Antoine Lavoisier: Science, Administration, and Revolution* (Oxford: Blackwell, 1993), pp. 25–30.

16. Louise Yvonne Palmer, "The Early Scientific Work of Antoine Laurent Lavoisier: In the Field and in the Laboratory, 1763–1767," Diss., Yale University, 1998.

17. Ibid., pp. 120–198.

18. Daniel P. Todes, *Pavlov's Physiology Factory: Experiment, Interpretation, Laboratory Enterprise* (Baltimore: Johns Hopkins University Press, 2002), pp. 50–63.

19. Frederic L. Holmes, "Seymour Benzer and the Definition of the Gene," in *The Concept of the Gene in Development and Evolution,* ed. Peter Beurton, Raphael Falk, and Hans-Jörg Rheinberger (Cambridge: Cambridge University Press, 2000), pp. 132–134, 140.

20. Elof Axel Carlson, *Genes, Radiation, and Society: The Life and Work of H. J. Muller* (Ithaca: Cornell University Press, 1981), pp. 57–90, quotations, pp. 59, 63.

21. Gerald L. Geison, *Michael Foster and the Cambridge School of Physiology* (Princeton: Princeton University Press, 1978).

CHAPTER 4. MASTERY OF A DOMAIN

1. Marcellin Berthelot, *La révolution chimique: Lavoisier* (Paris: Alcan, 1890), p. 3.
2. Jean Louis Faure, *Claude Bernard* (Paris: Crès, 1925), pp. iii, 16–17.
3. Howard Gardner, *Creating Minds: An Anatomy of Creativity* (New York: Basic Books, 1993), p. 32.
4. Frederic L. Holmes, *Antoine Lavoisier—the Next Crucial Year* (Princeton: Princeton University Press, 1998).
5. For some of the details, see Frederic L. Holmes, *Lavoisier and the Chemistry of Life* (Madison: University of Wisconsin Press, 1985), pp. 41–128.
6. Claude Bernard, *Principes de médecine expérimentale,* ed. Léon Delhoume (Paris: Presses Universitaires, 1947), p. 221.
7. Frederic L. Holmes, *Claude Bernard and Animal Chemistry* (Cambridge: Harvard University Press, 1974), pp. 197–444.
8. Frederic L. Holmes, *Hans Krebs: The Formation of a Scientific Life, 1900–1933* (New York: Oxford University Press, 1991), pp. 208–236.
9. Ibid., pp. 238–436.
10. Frederic L. Holmes, "Between Molecular Biology and the Classical Gene: The Pathway of Seymour Benzer into the rII Region," unpubl. manuscript, chap. 3.
11. Ibid., chaps. 3, 4.
12. Frederic L. Holmes, "Seymour Benzer and the Definition of the Gene," in *The Concept of the Gene in Development and Evolution,* ed. Peter Beurton, Raphael Falk, and Hans-Jörg Rheinberger (Cambridge: Cambridge University Press, 2000), pp. 115–155.
13. Frederic L. Holmes, *Meselson, Stahl, and the Replication of DNA: A History of "the Most Beautiful Experiment in Biology"* (New Haven: Yale University Press, 2001).
14. Nathaniel C. Comfort, *The Tangled Field: Barbara McClintock's Search for the Patterns of Genetic Control* (Cambridge: Harvard University Press, 2001), pp. 49–56.
15. Ibid., pp. 56–120.

CHAPTER 5. IS DISTINCTION ACHIEVED OR CONFERRED?

1. L. Pearce Williams, *Michael Faraday* (New York: Basic Books, 1964), pp. 171, 161, 185.
2. Augustine Brannigan, *The Social Basis of Scientific Discoveries* (Cambridge: Cambridge University Press, 1981), pp. 11, 82–83.
3. Robert K. Merton, "The Sociology of Science: An Episodic Memoir," in *The Sociology of Science in Europe,* ed. Robert K. Merton and Jerry Gaston (Carbondale: Southern Illinois University Press, 1977), pp. 89–93.
4. Frederic L. Holmes, *Antoine Lavoisier—the Next Crucial Year* (Princeton: Princeton University Press, 1998), pp. 124–136.
5. Ibid., pp. 128–129, 136–139.
6. Frederic L. Holmes, *Lavoisier and the Chemistry of Life: An Exploration of Scientific Creativity* (Madison: University of Wisconsin Press, 1985), p. 42.
7. Ibid., p. 43.
8. Holmes, *Next Crucial Year,* p. 118.

9. Ferdinando Abbri, *Le terre, l'acqua, le arie: La rivoluzione chimica del Settecento* (Bologna: Il Mulino, 1984), pp. 160–161.

10. Joseph Priestley, *Experiments and Observations on Different Kinds of Air,* vol. 2 (London: Johnson, 1775), pp. 121–122.

11. Robert K. Merton, with Harriet Zuckerman, "Age, Aging, and the Age Structure in Science," in Robert K. Merton, *The Sociology of Science: Theoretical and Empirical Investigations,* ed. Norman W. Storer (Chicago: University of Chicago Press, 1973), p. 510.

12. For biographical details, see J. M. D. Olmsted, *Claude Bernard, Physiologist* (New York: Harper and Brothers, 1938), pp. 30–38.

13. Ibid., pp. 30–31, 42–44.

14. Frederic L. Holmes, *Claude Bernard and Animal Chemistry* (Cambridge: Harvard University Press, 1974), p. 441.

15. Frederic L. Holmes, *Hans Krebs: The Formation of a Scientific Life, 1900–1933* (New York: Oxford University Press, 1991), pp. 209–276.

16. Ibid., pp. 349–384.

17. Ibid., pp. 385.

18. Frederic L. Holmes, *Hans Krebs: Architect of Intermediary Metabolism, 1933–1937* (New York: Oxford University Press, 1993), pp. 3–203.

19. Frederic L. Holmes, *Meselson, Stahl, and the Replication of DNA* (New Haven: Yale University Press, 2001), pp. 319–387.

CHAPTER 6. DILEMMAS OF THE AGING SCIENTIST

1. Robert K. Merton, with Harriet Zuckerman, "Age, Aging, and the Age Structure in Science," in Robert K. Merton, *The Sociology of Science: Theoretical and Empirical Investigations,* ed. Norman W. Storer (Chicago: University of Chicago Press, 1973), pp. 503, 510–511.

2. Ibid., p. 529.

3. Ibid., pp. 531–532.

4. Frederic L. Holmes, *Lavoisier and the Chemistry of Life* (Madison: University of Wisconsin Press, 1985), pp. 129–147.

5. Henry Guerlac, "Chemistry as a Branch of Physics," *Historical Studies in the Physical Sciences* 7 (1976): 193–276.

6. See among others, Carleton E. Perrin, "The Triumph of the Antiphlogistians," in *The Analytic Spirit: Essays in the History of Science in Honor of Henry Guerlac,* ed. Harry Woolf (Ithaca, N.Y.: Cornell University Press, 1981), pp. 40–63; Arthur Donovan, *Antoine Lavoisier: Science, Administration, and Revolution* (Oxford: Blackwell, 1993), pp. 157–187; and Bernadette Bensaude-Vincent, *Lavoisier: Mémoires d'une révolution* (Paris: Flammarion, 1993), pp. 234–312.

7. Holmes, *Lavoisier,* pp. 261–483.

8. Jan Golinski, *Science as Public Culture: Chemistry and Enlightenment in Britain, 1760–1820* (Cambridge: Cambridge University Press, 1992), p. 138.

9. For a comprehensive discussion of Bernard's discoveries and their historical place, see

J. M. D. Olmsted, *Claude Bernard, Physiologist* (New York: Harper Brothers, 1938), pp. 139–213.

10. Mirko D. Grmek, *Raisonnement expérimental et recherches toxicologiques chez Claude Bernard* (Paris: Droz, 1973), pp. 71–207.

11. Olmsted, *Bernard,* pp. 162–174; Frank G. Young, "Claude Bernard and the Glycogen Function of the Liver," *Annals of Science* 2 (1937): 47–83; Nikolaus Mani, *Die historischen Grundlagen der Leberforschung* (Basel: Schwabe, 1967), pp. 353–364.

12. Frederic L. Holmes, "Claude Bernard, the *milieu intérieur* and Regulatory Physiology," *History and Philosophy of the Life Sciences* 8 (1986): 3–25, esp. 22.

13. Hebbel H. Hoff and Roger Guillemin, "Claude Bernard and the Vasomotor System," in *Claude Bernard and Experimental Medicine,* ed. Francisco Grande and Maurice B. Visscher (Cambridge, Mass.: Schenkman, 1967), pp. 75–104.

14. Frederic L. Holmes, "Krebs, Hans Adolf," in *Dictionary of Scientific Biography,* ed. F. L. Holmes, vol. 17 (New York: Scribner's, 1990), pp. 496–506.

15. H. A. Krebs, "Control of Metabolic Processes," *Endeavour* 16 (1957): 125.

16. H. A. Krebs, "The History of the Tricarboxylic Acid Cycle," *Perspectives in Biology and Medicine* 14 (1970): 167–168.

17. Dean Keith Simonton, "Creative Expertise: A Life-Span Developmental Perspective," in *The Road to Excellence: The Acquisition of Expert Performance in the Arts and Sciences, Sports and Games,* ed. K. Anders Ericsson (Mahwah, N.J.: Erlbaum, 1996), pp. 227–253.

18. Ibid., pp. 234–235.

19. Olmsted, *Bernard,* pp. 67–84.

20. Ibid., pp. 117–118.

21. Jack E. Baldwin and Hans Krebs, "The Evolution of Metabolic Cycles," *Nature* 291 (1981): 381–382.

22. Ibid., p. 382.

23. Hans Krebs, F. L. Holmes, conversation, January 8, 1981.

CHAPTER 7. COMPLICATING THE PATHWAY METAPHOR

1. Howard E. Gruber, "The Evolving Systems Approach to Creative Work," in *Creative People at Work,* ed. Doris B. Wallace and Howard E. Gruber (New York: Oxford University Press, 1989), pp. 11, 13.

2. Ibid., p. 12.

3. David Gooding, *Experiment and the Making of Meaning* (Dordrecht: Kluwer, 1990), pp. 11–18.

4. Ibid., pp. 16, 18.

5. Ibid., pp. 135–161.

6. Gerald L. Geison, *The Private Science of Louis Pasteur* (Princeton: Princeton University Press, 1995), p. 37.

7. Ibid., pp. 95–96.

8. Daniel P. Todes, *Pavlov's Physiology Factory: Experiment, Interpretation, Laboratory Enterprise* (Baltimore: Johns Hopkins University Press, 2002), pp. 217–254.

9. Frederic L. Holmes, "Seymour Benzer and the Definition of the Gene," in *The Concept of the Gene in Development and Evolution,* ed. Peter Beurton, Raphael Falk, and Hans-Jörg Rheinberger (Cambridge: Cambridge University Press, 2000), pp. 115–155.

10. Jonathan Weiner, *Time, Love, Memory: A Great Biologist and His Quest for the Origins of Behavior* (New York: Knopf, 1999).

CHAPTER 8. COMPLICATING THE EPISODIC RHYTHMS

1. For details concerning this and the following paragraphs, see Frederic L. Holmes, *Lavoisier and the Chemistry of Life* (Madison: University of Wisconsin Press, 1985), pp. 201–238.

2. Ibid., pp. 264–269.

3. Ibid., pp. 270–289.

4. Ibid., pp. 277, 289.

5. Armand Seguin and A. L. Lavoisier, "Premier mémoire sur la respiration des animaux," *Mémoires de l'Académie des Sciences* (Paris, 1789 [publ. 1793]).

6. Holmes, *Lavoisier and the Chemistry of Life,* p. 19, omitting here phrases that Lavoisier crossed out.

7. Ibid., pp. 63–90.

8. Ibid., pp. 151–198.

9. Ibid., pp. 120–126.

10. Ibid., pp. 237–259.

11. Ibid., pp. 440–468.

12. Frederic L. Holmes, *Claude Bernard and Animal Chemistry* (Cambridge: Harvard University Press, 1974), pp. 197–376.

13. Ibid., pp. 278–297.

14. Ibid., pp. 377–400.

15. Ibid., pp. 401–444.

16. See the bibliography of his publications in Hans Krebs, with Anne Martin, *Reminiscences and Reflections* (Oxford: Oxford University Press, 1981), pp. 269–289. I have omitted here review articles and research publications based on projects carried out primarily by students, some of which were co-authored by Krebs.

17. Ibid.

18. Frederic L. Holmes, *Hans Krebs: The Formation of a Scientific Life, 1900–1933* (New York: Oxford University Press, 1991), pp. 247–341.

19. Ibid., pp. 307, 313–314.

20. Ibid., pp. 397–415.

21. Frederic L. Holmes, *Hans Krebs: Architect of Intermediary Metabolism, 1933–1937* (New York: Oxford University Press, 1993), pp. 10–39.

CHAPTER 9. PREDICTABILITY AND UNPREDICTABILITY

1. François Jacob, *La souris, la mouche et l'homme* (Paris, 1997), pp. 25–26.

2. Hans-Jörg Rheinberger, *Toward a History of Epistemic Things: Synthesizing Proteins in a Test Tube* (Stanford: Stanford University Press, 1997), pp. 27–28, 31–33, 34, 36, 67.

3. Ibid., pp. 38–54, 84–101, 114–132, 143–175, 197–203.

4. Howard E. Gruber, "The Evolving Systems Approach to Creative Work," in *Creative People at Work,* ed. Doris B. Wallace and Howard E. Gruber (New York: Oxford University Press, 1989), pp. 10–11.

5. In adapting Rheinberger's concept of experimental systems to her study of tobacco mosaic virus as a model system, Angela Creager similarly argues that "when looking to experimental systems as interacting elements in a broader scientific story, one must be able to speak to the decisions of scientists, who are the principal agents operating on and between different experimental systems." Angela N. H. Creager, *The Life of a Virus: Tobacco Mosaic Virus as an Experimental Model* (Chicago: University of Chicago Press, 2002), p. 327.

6. Lavoisier, "Sur la cause de laugmentation de pesanteur quacquierent les metaux et quelques autres substances par la calcination," reproduced in C. E. Perrin, "Lavoisier's Thoughts on Calcination and Combustion, 1772–1773," Isis 77 (1986): 662–665.

7. Frederic L. Holmes, *Antoine Lavoisier—the Next Crucial Year* (Princeton: Princeton University Press, 1998), pp. 15–23.

8. Ibid., pp. 30–40.

9. Ibid., pp. 104–124.

10. Frederic L. Holmes, *Lavoisier and the Chemistry of Life* (Madison: University of Wisconsin Press, 1985), pp. 44–48.

11. Ibid., pp. 49–128.

12. Claude Bernard, *An Introduction to the Study of Experimental Medicine* (New York: Collier Books, 1961), p. 156.

13. Ibid., pp. 152–155.

14. Ibid., pp. 165–168.

15. Ibid., pp. 168–170.

16. Frederic L. Holmes, *Claude Bernard and Animal Chemistry* (Cambridge: Harvard University Press, 1974), p. 179.

17. Bernard, *Introduction,* p. 163.

18. Holmes, *Claude Bernard,* pp. 422–423.

19. Mirko D. Grmek, *Le legs de Claude Bernard* (Paris: Fayard, 1997), pp. 207–249.

20. Holmes, *Claude Bernard,* pp. 423–444.

21. Frederic L. Holmes, *Hans Krebs: The Formation of a Scientific Life, 1900–1933* (New York: Oxford University Press, 1991), p. 406.

22. Ibid., pp. 416–417.

23. Ibid., p. 428.

24. Frederic L. Holmes, *Hans Krebs: Architect of Intermediary Metabolism, 1933–1937* (New York: Oxford University Press, 1993), p. 174.

25. Holmes, *Krebs* (1991), p. 250.

26. Rheinberger, *Epistemic Things,* pp. 78–80.

27. Seymour Benzer, personal conversation, January 5, 2002.

28. M. Delbrück and S. E. Luria, "Interference between Bacterial Viruses: I. Interference between Two Viral Particles Acting upon the Same Host, and the Mechanism of Viral Growth," *Archives of Biochemistry* 1 (1943): 111–141.

29. Frederic L. Holmes, *Meselson, Stahl, and the Replication of DNA: A History of "the Most Beautiful Experiment in Biology"* (New Haven: Yale University Press, 2001), pp. 168–178.
30. Ibid., pp. 185–186.
31. Ibid., pp. 194–195.
32. Ibid., p. 197.
33. Ibid., pp. 197–229, 287–321.
34. Frederic L. Holmes, "Seymour Benzer and the Convergence of Molecular Biology with Classical Genetics," in *Mapping the Gene*, ed. Jean-Paul Gaudilliere and Hans-Jörg Rheinberger, forthcoming.
35. Bruno Latour and Steve Woolgar, *Laboratory Life: The Social Construction of Scientific Facts* (Beverly Hills: Sage, 1979), pp. 244–252.
36. Thomas S. Kuhn, *The Structure of Scientific Revolutions,* 2d ed. (Chicago: University of Chicago Press, 1970), pp. 52–53.
37. Ibid., p. 58.
38. Frederic L. Holmes, "The 'Revolution in Chemistry and Physics': Overthrow of a Reigning Paradigm or Competition between Contemporary Research Programs?" *Isis* 91 (2000): 735.

CHAPTER 10. INTERPLAY BETWEEN THOUGHT AND OPERATION

1. H. A. Simon, *Models of Discovery* (Dordrecht: Reidel, 1977), p. 286.
2. Howard E. Gruber, *Darwin on Man: A Psychological Study of Scientific Creativity,* 2d ed. (Chicago: University of Chicago Press, 1981), p. 10.
3. This and succeeding paragraphs follow the account in Frederic L. Holmes, *Lavoisier and the Chemistry of Life* (Madison: University of Wisconsin Press, 1985).
4. Ibid., pp. 320–324.
5. Ibid., p. 334.
6. This and the following paragraphs follow the narrative related in fuller detail in Frederic L. Holmes, *Hans Krebs: Architect of Intermediary Metabolism, 1933–1937* (New York: Oxford University Press, 1993), pp. 238–256.
7. Frederic L. Holmes, *Meselson, Stahl, and the Replication of DNA: A History of "the Most Beautiful Experiment in Biology"* (New Haven: Yale University Press, 2001), pp. 195–206.
8. This and the following paragraphs follow the narrative related in fuller detail in ibid., pp. 215–229.

CHAPTER 11. FLASHES OF INSIGHT, MOMENTS OF DISCOVERY

1. For an analysis of the relation between Kekulé's dream and the development of his structural formulas, see Alan J. Rocke, "Subatomic Speculations and the Origins of Structure Theory," *Ambix* 30 (1983): 1–18; for Kekulé's other well-known dream leading to the benzene ring, see Rocke, "Hypothesis and Experiment in the Early Development of Kekulé's Benzene Theory," *Annals of Science* 42 (1985): 355–381.
2. R. Taton, *Reason and Chance in Scientific Discovery* (New York: Science Editions, 1962), p. 21.

3. Charles Darwin, *The Autobiography of Charles Darwin and Selected Letters,* ed. Francis Darwin (New York: Dover, 1958), pp. 42–43.

4. Taton, *Reason and Chance,* p. 74.

5. Thomas S. Kuhn, *The Structure of Scientific Revolutions,* 2d ed. (Chicago: University of Chicago Press, 1970), pp. 89–90.

6. Taton, *Reason and Chance,* pp. 22–23, 28–29, 36.

7. Howard E. Gruber, "On the Relation between 'Aha Experiences' and the Construction of Ideas," *History of Science* 19 (1981): 41–59.

8. Ibid., pp. 42–43. See also Howard E. Gruber, *Darwin on Man: A Psychological Study of Scientific Creativity,* 2d ed. (Chicago: University of Chicago Press, 1981), pp. 150–174.

9. Kuhn, *Structure,* p. 118.

10. Larry Laudan, *Science and Values: The Aims of Science and Their Role in Scientific Debate* (Berkeley: University of California Press, 1984), pp. 62–102.

11. Frederic L. Holmes, *Lavoisier and the Chemistry of Life* (Madison: University of Wisconsin Press, 1985), pp. 118–120.

12. Ibid., p. 19.

13. Frederic L. Holmes, *Claude Bernard and Animal Chemistry* (Cambridge: Harvard University Press, 1974), pp. 261–262.

14. James D. Watson, *The Double Helix: A Personal Account of the Discovery of the Structure of DNA* (New York: Atheneum, 1968), pp. 178–196, quotation pp. 195–196.

15. Niels K. Jerne, "The Natural Selection Theory of Antibody Formation: Ten Years Later," in *Phage and the Origins of Molecular Biology,* ed. John Cairns, Gunther S. Stent, and James D. Watson (Cold Spring Harbor, N.Y.: Cold Spring Harbor Laboratory of Quantitative Biology, 1966), pp. 301–302.

16. Thomas Söderqvist, "The 'Onlie Begetter': Niels K. Jerne and the Origin of the Selection Theory of Antibody Formation," unpublished manuscript.

17. Frederic L. Holmes, *Hans Krebs: The Formation of a Scientific Life, 1900–1933* (New York: Oxford University Press, 1991), pp. 287–328.

18. Frederic L. Holmes, *Hans Krebs: Architect of Intermediary Metabolism, 1933–1937* (New York: Oxford University Press, 1993), pp. 349, 396–403.

19. Holmes, *Krebs* (1991), p. 327.

20. Seymour Benzer, "Adventures in the rII Region," in *Phage and the Origins,* pp. 157–165, quotation p. 161.

21. Frederic L. Holmes, "Seymour Benzer and the Convergence of Molecular Biology with Classical Genetics," forthcoming.

22. Gruber, "'Aha Experiences,'" p. 41.

23. Barbara McClintock had "many experiences" that she described as "getting signals from my subconscious that I cannot tell you necessarily where they come from, but the whole thing is solved suddenly." Yet she described the process not as a holistic leap, but a rapid, rational, computational process, one of orienting herself, thinking very hard about a problem, and "integrating properly." Nathaniel C. Comfort, *The Tangled Field: Barbara McClintock's Search for the Patterns of Genetic Control* (Cambridge: Harvard University Press, 2001), pp. 67–68.

CONCLUSION

1. Robert J. Richards, *Darwin and the Emergence of Evolutionary Theories of Mind and Behavior* (Chicago: University of Chicago Press, 1987), pp. 17–19.
2. Hans-Jörg Rheinberger, *Toward a History of Epistemic Things: Synthesizing Proteins in the Test Tube* (Stanford: Stanford University Press, 1997), p. 74.

AFTERWORD

1. Details concerning relevant aspects of the history of wave optics can be found in J. Z. Buchwald, *The Rise of the Wave Theory of Light: Optical Theory and Experiment in the Early Nineteenth Century* (Chicago: University of Chicago Press, 1989).
2. On which see J. Z. Buchwald, *The Creation of Scientific Effects: Heinrich Hertz and Electric Waves* (Chicago: University of Chicago Press, 1994).
3. Frederic L. Holmes, *Lavoisier and the Chemistry of Life: An Exploration of Scientific Creativity* (Madison: University of Wisconsin Press, 1985), p. 192.
4. Frederic L. Holmes, *Eighteenth-century Chemistry as an Investigative Enterprise* (Berkeley, Office for History of Science and Technology, University of California at Berkeley, 1989), p. 126.

Index